MACMILLAN
WORK OUT
SERIES

Work Out

Pure Mathematics

'A' Level

The titles in this series

MACMILLAN
WORK OUT
SERIES

Work Out

Pure Mathematics

'A' Level

B. Haines

and

R. Haines

MACMILLAN

First published 1986
Reprinted 1986, 1987, 1989

Published by
MACMILLAN EDUCATION LTD
Houndmills, Basingstoke, Hampshire RG21 2XS
and London
Companies and representatives
throughout the world

Typesetting and Illustration by TecSet Ltd,
Sutton, Surrey
Printed in Hong Kong

British Library Cataloguing in Publication Data
Haines, B.
Work out pure mathematics 'A' level.—
(Macmillan work out series).—(Macmillan master series)
1. Mathematics—Examinations, questions, etc.
I. Title   II. Haines, R.
510'.76   QA43
ISBN 0–333–37398–7
ISBN 0–333–44109–5 Export pbk

To our Parents

Contents

Acknowledgements

Over many years the questions set by the various examination boards have stimulated and enhanced the teaching of mathematics throughout education. Everyone involved in mathematics, both the teachers and the taught, owes a debt to the boards for the ever-present challenge that new examination questions bring to mathematics education.

Our thanks go to everyone who has helped with the preparation of this book, especially to Nicholas, who, as word processor and checker, has made numerous suggestions and corrections.

We would be greatly indebted to anyone notifying us of any errors.

The University of London Entrance and School Examinations Council accepts no responsibility whatsoever for the accuracy or method in the answers given in this book to actual questions set by the London Board.

Acknowledgement is made to the Southern Universities' Joint Board for School Examinations for permission to use questions taken from their past papers but the Board is in no way responsible for answers that may be provided and they are solely the responsibility of the authors.

The Associated Examining Board, the University of Oxford Delegacy of Local Examinations, the Northern Ireland Schools Examination Council and the Scottish Examination Board wish to point out that worked examples included in the text are entirely the responsibility of the authors and have neither been provided nor approved by the Board.

Examination Boards for 'A' Level

Syllabuses and past examination papers can be obtained from:

The Associated Examining Board (AEB)
Stag Hill House Guildford Surrey GU2 5XJ

University of Cambridge Local Examinations Syndicate (UCLES)
Syndicate Buildings Hills Road Cambridge CB1 2EU

Joint Matriculation Board (JMB)
78 Park Road Altrincham Ches. WA14 5QQ

University of London Schools Examinations Board (L)
University of London Publications Office 52 Gordon Square London WC1E 6EE

University of Oxford (OLE)
Delegacy of Local Examinations Ewert Place Summertown Oxford OX2 7BZ

Oxford and Cambridge Schools Examination Board (O&C)
10 Trumpington Street Cambridge BC2 1QB

Scottish Examination Board (SEB)
Robert Gibson & Sons (Glasgow) Ltd 17 Fitzroy Place Glasgow G3 7SF

Southern Universities' Joint Board   **(SUJB)**
Cotham Road Bristol BS6 6DD

Welsh Joint Education Committee   **(WJEC)**
245 Western Avenue Cardiff CF5 2YX

Northern Ireland Schools Examination Council   **(NISEC)**
Examinations Office Beechill House Beechill Road Belfast BT8 4RS

Introduction

This Work Out is not just another text book. It is based on the 'A' level Common Core Mathematics syllabus and has been designed to help students to obtain the best possible grades in their 'A' level examination. In addition it has been written to provide a bridge for students, especially BTEC students, wishing to go into higher education. Students entering higher education often find they are unfamiliar with some 'A' level topics and lack sufficient practice in topics which the lecturers assume to be known.

Each chapter in the book starts with a brief list of formulae and is followed by many 'A' level type questions, each with a complete solution. At the end of each chapter more 'A' level questions are set as exercises with the important stages in the working of each one being given to facilitate easy reference.

How to Use the Book

(a) By repeated use and practice, endeavour to become familiar with the frequently used formulae listed in the fact sheets.

(b) Practice in answering examination questions is important. Open the book at a definite topic, choose a question and cover up the solution until you have tried to do it by yourself. If you get really stuck, your mind will be receptive when you uncover the solution.

(c) When answering multiple-choice questions, work through your solution in the same way as you would an ordinary question. If you are unable to answer the question, substituting the suggested answers into the question can often help to eliminate some solutions. Only guess an answer as a last resort, and not even then if the marking scheme penalizes wrong answers.

Revision

(a) Your educational establishment should be able to supply you with a syllabus and typical examination papers. Failing that, you should write to the secretary of the board whose examination you plan to take. A list of addresses is given on pages vii and viii.

(b) Use the book to revise topics you have already covered, before trying complete papers.

(c) Frequent reference to the formula booklet in examinations wastes precious time and increases the risk of losing your train of thought in the middle of a solution, while, as often as not, the formula you seek is not listed. Familiarize yourself with the contents of your formula sheet before the examination!

(d) Development of an examination technique is very important and may take months to evolve. Well before the examination, may be before you finish the syllabus, choose a convenient time and do a paper 'to time' under examination conditions (quiet!). You may well be appalled at how badly you do. Go through the paper again as soon as possible, referring to similar questions in this book. At this time your mind will be receptive to ways of overcoming the difficulties you have encountered. If you are able to do one paper 'to time' each fortnight for two or three months you will develop a sixth sense which enables you to spot 'easy' questions.

(e) Get into the habit of writing solutions tidily *first time*. So many students are content with scruffy solutions during the months leading up to the examination. Perhaps they expect a fairy to wave a wand and change their untidiness into a perfect script on the day! Unfortunately there is a shortage of such fairies, so get into the 'neat first time' habit.

The Examination

(a) Do the 'easy' questions at the start of the examination; this boosts your confidence.

(b) If you get stuck on a question and cannot see an alternative approach, cut your losses and go to another question. Examinations have been failed by good candidates spending too long stuck on one or two questions.

(c) If you find you are running out of time then look for questions which give 'part-way' answers, usually after asking bookwork. Quite often the problem part of the question is a straightforward application of the bookwork, is easy to do and is rewarding in marks. Do it, even if you cannot do the bookwork.

(d) Never cross out. You may be crossing out marks.

(e) Never walk out of an examination. Re-read the questions right through; even if you cannot do the first parts of the questions there will be some parts of some questions you can try.

(f) Good luck!

1 Polynomials and Indices

Algebraic operations on polynomials and rational functions. Factors of polynomials. The remainder theorem. The factor theorem. Positive and negative rational indices.

1.1 Fact Sheet

(a) Long Division

If a polynomial $f(x)$ is divided by a polynomial $g(x)$, of a lower degree than $f(x)$, the remainder is a polynomial of a lower degree than $g(x)$.

Example: If $x^5 - 3x^2 + x - 2$ (degree 5) is divided by $x^2 + 2x - 5$ (degree 2) the remainder will be of a lower degree than $x^2 + 2x - 5$, that is $ax + b$ (degree 1, or degree 0 if $a = 0$).

(b) Remainder Theorem

When a polynomial $f(x)$ is divided by $(x - a)$ the remainder is $f(a)$.
$f(x) = (x - a)g(x) + f(a)$.

(c) Factor Theorem

- If $f(a) = 0$ then $(x - a)$ is a factor of $f(x)$.
- If $f(a) = 0$ and $f'(a) = 0$ then $(x - a)$ is a repeated factor of $f(x)$.

(d) Identities

If $f(x) \equiv g(x)$ then $f(a) = g(a)$ for all a
and the coefficient of x^n in $f(x)$ = coefficient of x^n in $g(x)$ for all n.

(e) Indices

- $a^m \times a^n = a^{m+n}$.
- $a^{-m} = 1/a^m$.
- $a^{m/n} = \sqrt[n]{a^m}$.

- $a^m/a^n = a^{m-n}$.
- $a^0 = 1$.
- $(a^m)^n = a^{mn} = (a^n)^m$.

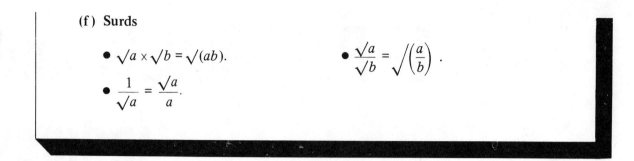

(f) Surds

- $\sqrt{a} \times \sqrt{b} = \sqrt{(ab)}$.
- $\dfrac{1}{\sqrt{a}} = \dfrac{\sqrt{a}}{a}$.

- $\dfrac{\sqrt{a}}{\sqrt{b}} = \sqrt{\left(\dfrac{a}{b}\right)}$.

1.2 Worked Examples

1.1
(a) If $f(x) = 6x^2 - 7x - 3$ find the values of $f(0)$, $f(2)$, $f(-1)$.
(b) If $f(x) = 2x^3 - 7x^2 + 4x - 3$ prove that $7f(1) = 4f(2)$.

- (a) $f(0) = 6(0)^2 - 7(0) - 3 = -3$,
  $f(2) = 6(2)^2 - 7(2) - 3 = 24 - 14 - 3 = 7$,
  $f(-1) = 6(-1)^2 - 7(-1) - 3 = 6 + 7 - 3 = 10$.

  (b) $f(1) = 2(1)^3 - 7(1)^2 + 4(1) - 3 = 2 - 7 + 4 - 3 = -4$
  $$7f(1) = -28.$$
  $f(2) = 2(2)^3 - 7(2)^2 + 4(2) - 3 = 16 - 28 + 8 - 3 = -7$
  $$4f(2) = -28.$$
  Hence $7f(1) = 4f(2)$.

1.2 Find a and b if $a(x + 3)^2 + b(x - 2) + 1 \equiv 3x^2 + 20x + 24$.

- **Substitute values for x.**

  When $x = -3$, $\quad -5b + 1 = 3(-3)^2 + 20(-3) + 24$
  $$= -9$$
  $$\Rightarrow b = 2.$$
  When $x = 2$, $\quad 25a + 1 = 3(2)^2 + 20(2) + 24$
  $$= 76$$
  $$\Rightarrow a = 3.$$

Alternatively, expand the left-hand side of this identity:

l.h.s. $\quad a(x + 3)^2 + b(x - 2) + 1 \equiv a(x^2 + 6x + 9) + bx - 2b + 1$
$$\equiv ax^2 + 6ax + 9a + bx - 2b + 1$$
$$\equiv ax^2 + x(6a + b) + 9a - 2b + 1.$$
If $ax^2 + x(6a + b) + 9a - 2b + 1 \equiv 3x^2 + 20x + 24$ then the coefficients of like terms may be equated.

That is, $\quad a = 3 \qquad 6a + b = 20 \qquad 9a - 2b + 1 = 24$.

Solve the first two equations for a and b, and check in the third equation.

Hence $\qquad\qquad\qquad a = 3, \qquad b = 2$.

1.3 Given that $(x + 1)$ and $(x - 2)$ are factors of $x^4 - 3x^3 + ax^2 + bx + 4$, find the values of a and b.

- If $(x + 1)$ and $(x - 2)$ are factors of $f(x)$ then $f(-1)$ and $f(2)$ are zero.
  $f(-1) = (-1)^4 - 3(-1)^3 + a(-1)^2 + b(-1) + 4 = 0$,

  i.e. $\qquad\qquad\qquad\qquad\qquad 8 + a - b = 0.$ (1)

  $f(2) = (2)^4 - 3(2)^3 + a(2)^2 + b(2) + 4 = 0$,

  i.e. $\qquad\qquad\qquad\qquad\qquad 4a + 2b - 4 = 0,$ (2)

  i.e. $\qquad\qquad\qquad\qquad\qquad 2a + b - 2 = 0.$ (3)

  Adding equations 1 and 3:
  $$3a + 6 = 0, \qquad \text{i.e.} \quad a = -2.$$
  Substitute into equation 1:
  $$8 - 2 - b = 0, \qquad \text{i.e.} \quad b = 6.$$

1.4 The polynomial $x^3 + 2x^2 + ax + b$, where a, b are constants, leaves a remainder of 7 and 17 when divided by $(x - 2)$ and $(x + 3)$ respectively. Find the values of a and b and the remainder when this polynomial is divided by $(x - 4)$.

- Let $g(x) = x^3 + 2x^2 + ax + b$,
  $g(2) = $ the remainder when $g(x)$ is divided by $(x - 2)$.
  $g(2) = 8 + 8 + 2a + b = 7$,

  i.e. $\qquad\qquad\qquad\qquad\qquad 2a + b = -9$ (1)

  Similarly, $g(-3) = -27 + 18 - 3a + b = 17$,

  i.e. $\qquad\qquad\qquad\qquad\qquad -3a + b = 26$ (2)

  Subtracting equation 2 from equation 1:
  $$5a = -35, \qquad a = -7.$$
  Substituting into equation 1:
  $$-14 + b = -9, \qquad b = 5.$$
  Therefore $\qquad\qquad\qquad\qquad g(x) = x^3 + 2x^2 - 7x + 5,$
  $$g(4) = 64 + 32 - 28 + 5 = 73.$$
  The remainder when $g(x)$ is divided by $(x - 4)$ is 73.

1.5 Given that $g(x) = ax^3 + x^2 + bx - 2$, and $g(2)$ is $6b - 2a + 2$, then $a : b$ is

A, 3 : 4;   B, 5 : 2;   C, 3 : 2;   D, 2 : 5;   E, 4 : 5.

- If $g(x) = ax^3 + x^2 + bx - 2$ then $g(2) = 8a + 4 + 2b - 2$
  $$= 8a + 2b + 2.$$
  Hence $6b - 2a + 2 = 8a + 2b + 2 \quad \Rightarrow \quad 4b = 10a.$
  Hence $a/b = 4/10, \qquad a : b = 2 : 5.$ **Answer D**

1.6 Show that $17(1 - 1/17^2)^{1/2} = n\sqrt{2}$ where n is an integer.

- **Remember:** $a\sqrt{b} = \sqrt{(a^2 b)}$.

3

$$17 \left(1 - \frac{1}{17^2}\right)^{1/2} = \left[(17)^2 \left(1 - \frac{1}{17^2}\right)\right]^{1/2} = (17^2 - 1)^{1/2} = (288)^{1/2}.$$

$288 = 2 \times 144 = 2 \times 12^2 \quad \Rightarrow \quad (288)^{1/2} = 2^{1/2} \times 12,$
hence $17(1 - 1/17^2)^{1/2} = 12\sqrt{2}.$

1.7 When a polynomial $g(x)$ is divided by $(x + 3)$ the remainder is 8, and when $g(x)$ is divided by $(x - 2)$ the remainder is 3. Find the remainder when $g(x)$ is divided by $(x - 2)(x + 3)$.

● **When a polynomial is divided by $(x - 2)$ $(x + 3)$ (by a quadratic), the remainder is of the form $ax + b$ (linear).**

Let $g(x) = (x - 2)(x + 3) f(x) + ax + b.$

Then $g(-3) = 0 + -3a + b = 8 \quad \Rightarrow \quad -3a + b = 8$ (1)

$\qquad g(2) = 0 + 2a + b = 3 \quad \Rightarrow \quad 2a + b = 3.$ (2)

Subtracting equation 1 from equation 2,

$$5a = -5, \qquad a = -1.$$

Substituting in equation 2,

$$-2 + b = 3, \qquad b = 5.$$

The remainder when $g(x)$ is divided by $(x - 2)(x + 3)$ is $-x + 5.$

1.8 Use the factor theorem to find a linear factor of $P(x)$, where

$$P(x) = x^3 - 3x^2 - 10x + 24.$$

Hence express $P(x)$ as a product of three linear factors.

● **If a factor of $P(x)$ is of the form $(x - a)$ then a must be a factor of 24. Try the smallest factors first!**

$P(1) = 1 - 3 - 10 + 24 \neq 0 \qquad (x - 1)$ is not a factor.
$P(-1) = -1 - 3 + 10 + 24 \neq 0 \qquad (x + 1)$ is not a factor.
$P(2) = 8 - 12 - 20 + 24 = 0 \qquad (x - 2)$ is a factor.

At this stage either persevere with other possible factors or carry out a division by $(x - 2)$.

$$
\begin{array}{r}
x^2 - x - 12 \\
x - 2 \overline{\smash{)}\, x^3 - 3x^2 - 10x + 24} \\
\underline{x^3 - 2x^2} \\
- x^2 - 10x \\
\underline{- x^2 + 2x} \\
- 12x + 24 \\
\underline{- 12x + 24.}
\end{array}
$$

$x^2 - x - 12$ has factors $(x + 3)(x - 4)$ by inspection,
hence $P(x) = (x - 2)(x^2 - x - 12)$
$\qquad\qquad\quad = (x - 2)(x + 3)(x - 4).$

1.9 Given that $f(x) = 2x^4 + ax^3 + bx^2 - 8x + c$, find the real coefficients a, b and c when the following conditions are satisfied:

(a) $(x + 2)$ is a factor of $f(x)$ and $f'(x)$;

(b) when $f(x)$ is divided by $(x - 2)$ the remainder is 16.

Factorize $f(x)$ completely.

- **If $(x + 2)$ is a factor of $f(x)$ and $f'(x)$ then $(x + 2)^2$ is a factor of $f(x)$.**

$f(-2) = 0$ and $f'(-2) = 0$

$f(x) = 2x^4 + ax^3 + bx^2 - 8x + c, \qquad f'(x) = 8x^3 + 3ax^2 + 2bx - 8,$

$f(-2) = 32 - 8a + 4b + 16 + c = 0 \quad \Rightarrow \quad 8a - 4b - c = 48, \qquad (1)$

$f'(-2) = -64 + 12a - 4b - 8 = 0 \quad \Rightarrow \quad 12a - 4b = 72, \qquad (2)$

$f(2) = 32 + 8a + 4b - 16 + c = 16 \quad \Rightarrow \quad 8a + 4b + c = 0. \qquad (3)$

Adding equations 1 and 3: $16a = 48,$ $a = 3.$

Substitute into equation 2: $36 - 4b = 72,$ $b = -9.$

Substitute into equation 3: $24 - 36 + c = 0,$ $c = 12.$

Hence $f(x) = 2x^4 + 3x^3 - 9x^2 - 8x + 12 = (x + 2)^2 g(x).$

$$
\begin{array}{r}
2x^2 - 5x + 3 \\
x^2 + 4x + 4 \overline{)\, 2x^4 + 3x^3 - 9x^2 - 8x + 12} \\
\underline{2x^4 + 8x^3 + 8x^2} \\
-5x^3 - 17x^2 - 8x \\
\underline{-5x^3 - 20x^2 - 20x} \\
3x^2 + 12x + 12 \\
\underline{3x^2 + 12x + 12.}
\end{array}
$$

$f(x) = (x + 2)^2 (2x^2 - 5x + 3) = (x + 2)^2 (x - 1)(2x - 3).$

1.10 Simplify, without the use of tables or calculator,

$$\frac{\sqrt{3} + \sqrt{12} + \sqrt{108} - \sqrt{75}}{\sqrt{6} - \sqrt{96} + \sqrt{150}}.$$

- **Remember that $\sqrt{ab} = \sqrt{a}\sqrt{b}$ and $a\sqrt{c} + b\sqrt{c} = (a + b)\sqrt{c}$.**

$\sqrt{12} = \sqrt{(4 \times 3)} = \sqrt{4} \times \sqrt{3} = 2\sqrt{3}; \qquad \sqrt{108} = \sqrt{36} \times \sqrt{3} = 6\sqrt{3};$

$\sqrt{75} = \sqrt{25} \times \sqrt{3} = 5\sqrt{3}; \qquad \sqrt{96} = \sqrt{16} \times \sqrt{6} = 4\sqrt{6};$

$\sqrt{150} = \sqrt{25} \times \sqrt{6} = 5\sqrt{6}.$

Hence $\dfrac{\sqrt{3} + \sqrt{12} + \sqrt{108} - \sqrt{75}}{\sqrt{6} - \sqrt{96} + \sqrt{150}} = \dfrac{\sqrt{3} + 2\sqrt{3} + 6\sqrt{3} - 5\sqrt{3}}{\sqrt{6} - 4\sqrt{6} + 5\sqrt{6}}$

$$= \frac{4\sqrt{3}}{2\sqrt{6}} = \frac{2}{\sqrt{2}}$$

$$= \sqrt{2}.$$

1.11 Find, without using tables or calculator, the value of x, given that

$$\frac{2^{3x+7}}{4^{2x-2}} = \frac{8^{x-3}}{32^{5-x}}.$$

- Notice that all the base numbers are powers of 2. Express each term as a power of 2.

$$4^{2x-2} = (2^2)^{2x-2} = 2^{4x-4}; \qquad 8^{x-3} = (2^3)^{x-3} = 2^{3x-9};$$
$$32^{5-x} = (2^5)^{5-x} = 2^{25-5x}.$$

Hence

$$\frac{2^{3x+7}}{2^{4x-4}} = \frac{2^{3x-9}}{2^{25-5x}}.$$

Using the rules of indices, $2^a/2^b = 2^{a-b}$,

i.e. $2^{(3x+7)-(4x-4)} = 2^{(3x-9)-(25-5x)}$,

i.e. $2^{-x+11} = 2^{8x-34} \quad \Rightarrow \quad -x + 11 = 8x - 34, \quad x = 5.$

1.3 Exercises

1.1 Given that $(x + 3)$ is a factor of f(x) where f$(x) \equiv 2x^3 - ax + 12$, find the constant a. Express f(x) as a product of linear factors. (L)

1.2 Find the quotient and the remainder when the polynomial $x^2(x - 1)(2x + 3)$ is divided by the polynomial $x^2 + 2$.

1.3 If f(x) is a polynomial in x, show that when f(x) is divided by $(x - a)$ the remainder is f(a).

When $x^3 + ax^2 + bx + c$ is divided by $(x - 3)$ the remainder is 18 and when divided by $x^2 - 3x + 2$ the remainder is 10. Find the values of a, b and c.

1.4 Find, without using tables or calculator, the exact value of

$$\frac{(2+\sqrt{3})^2}{2-\sqrt{3}} + \frac{(2-\sqrt{3})^2}{2+\sqrt{3}}.$$

1.5 The units digit in the answer to

$$123^4 - 421^5 + 932^3$$

is: A, 2; B, 7; C, 3; D, 8; E, 9.

1.6 Given that $2^{x+1} - 5^y = 131$, $2^{x-4} + 5^{y-2} = 13$, find x and y.

1.7 Given that f$(x) = 4x^4 + 12x^3 - 5x^2 - 21x + 10$, find by trial two integer solutions of the equation f$(x) = 0$. Hence factorise f(x) and solve the equation completely.

1.8 If $2x - 1$ is a factor of $2x^3 + bx^2 - 8x + 2$, b is equal to:

A, 9;   B, -5;   C, 1;   D, 3;   E, 7.

1.9 Find, without the use of tables or a calculator, the value of x, given that

$$\frac{2^{x+5}}{8^x} = \frac{4^{x-1}}{2^{2x-1}}.$$

1.10 Using the remainder theorem, or otherwise, show that $x - a - b - c$ is a factor of

$$(x - a)(x - b)(x - c) - (b + c)(c + a)(a + b).$$

Hence, or otherwise, solve the equation

$$(x - 2)(x + 3)(x + 1) - 4 = 0.$$

1.4 Outline Solutions to Exercises

1.1 $(x + 3)$ is a factor so

$$f(-3) = 0 \quad \Rightarrow \quad a = 14 \quad \Rightarrow \quad f(x) = 2x^3 - 14x + 12.$$

Trying integers gives $f(2) = 0$ and $f(1) = 0$.

$$f(x) = 2(x + 3)(x - 2)(x - 1).$$

1.2 $x^2(x - 1)(2x + 3) = 2x^4 + x^3 - 3x^2$.

$$
\begin{array}{r}
2x^2 + x - 7 \\
x^2 + 2 \,\overline{\big)\, 2x^4 + x^3 - 3x^2} \\
\underline{2x^4 \qquad + 4x^2} \\
x^3 - 7x^2 \\
\underline{x^3 \qquad + 2x} \\
-7x^2 - 2x \\
\underline{-7x^2 \qquad - 14} \\
-2x + 14
\end{array}
$$

Remainder

Quotient: $2x^2 + x - 7$;   remainder: $-2x + 14$.

1.3 Let $f(x) = (x - a)\, g(x) + r$.
Put $x = a$, so $f(a) = r$.

$$f(3) = 27 + 9a + 3b + c = 18 \text{ (given).} \tag{1}$$

$$x^2 - 3x + 2 = (x - 1)(x - 2).$$

Dividing $(x - 1)$ and $(x - 2)$ into $x^3 + ax^2 + bx + c$ give remainders of 10.

$$1 + a + b + c = 10 \tag{2}$$

and

$$8 + 4a + 2b + c = 10. \tag{3}$$

Solving these equations gives $a = -2$, $b = -1$, $c = 12$.

1.4 Put over common denominator: $(2 - \sqrt{3})(2 + \sqrt{3}) = 4 - 3 = 1$.

Expand numerator: $(2 + \sqrt{3})^3 + (2 - \sqrt{3})^3 = 52$.

1.5 $\qquad\qquad 3^4 = 81, \qquad 1^5 = 1, \qquad 2^3 = 8.$

Units digit: $\qquad\qquad\qquad\qquad 1 - 1 + 8 = 8.$ Answer **D**

1.6 Multiply second equation by 5^2 and add them:

$$2^{x+1} + 5^2 \cdot 2^{x-4} = 456 \quad \Rightarrow \quad 2^{x-4}(2^5 + 5^2) = 456$$
$$\Rightarrow \quad 2^{x-4} = 8 = 2^3.$$

Hence $\qquad\qquad\qquad\qquad (x - 4) = 3 \quad \Rightarrow \quad x = 7.$

Substitute into first equation:

$$2^8 - 5^y = 131, \qquad 5^y = 125, \qquad y = 3.$$
$$x = 7, y = 3.$$

1.7 Try integers. $f(1) = 0$, $f(-2) = 0$ so $x = 1$, $x = -2$ are solutions.

$$(x - 1)(x + 2) = x^2 + x - 2.$$

Long division gives:

$$f(x) = (x^2 + x - 2)(4x^2 + 8x - 5).$$

$4x^2 + 8x - 5 = 0$ has solutions $x = \frac{1}{2}$ or $x = -\frac{5}{2}$.
Hence roots $x = 1, -2, \frac{1}{2}, -\frac{5}{2}$.

1.8 Since $(2x - 1)$ is a factor, $f(\frac{1}{2}) = 0$.
Substitution of $x = \frac{1}{2}$ gives $b = 7$. Answer **E**

1.9 Remember that $8^x = 2^{3x}$ and $4^{x-1} = 2^{2x-2}$.
Equation simplifies to $2^{x+5-3x} = 2^{2x-2-2x+1}$.
Equating indices gives $x = 3$.

1.10 $f(x) = (x - a)(x - b)(x - c) - (b + c)(c + a)(a + b)$.
Putting $x = a + b + c$ gives $f(a + b + c) = 0$

$$\Rightarrow x - a - b - c \text{ is a factor.}$$

Let $a = 2, b = -3, c = -1$.
Then $x + 2$ is a factor of $(x - 2)(x + 3)(x + 1) - 4 = 0$.
Factorizing gives $(x + 2)(x^2 - 5) = 0$.

$$x = -2 \quad \text{or} \quad \pm\sqrt{5}.$$

2 Quadratic Functions

The general quadratic function in one variable including solution of quadratic equations, sketching graphs and finding maxima and minima by completing the square.

2.1 Fact Sheet

(a) The **quadratic equation** $ax^2 + bx + c = 0$ has two roots given by

$$x = \frac{-b \pm \sqrt{(b^2 - 4ac)}}{2a} \, .$$

(b) The **discriminant** $b^2 - 4ac$ gives the nature of the roots.

- If $b^2 - 4ac > 0$, roots are real and distinct.
- If $b^2 - 4ac = 0$, roots are real and equal.
- If $b^2 - 4ac < 0$, roots are complex.
- If $b^2 - 4ac \geqslant 0$ and $ac > 0$, roots have the same sign.
- The quadratic function $f(x) = ax^2 + bx + C > 0$ for all real x if $b^2 - 4ac < 0$ and $a > 0$.

(c) **Roots**

- If the roots of $ax^2 + bx + c = 0$ are α and β, then:

$$\alpha + \beta = -\frac{b}{a} \quad \text{and} \quad \alpha\beta = \frac{c}{a} \, .$$

- Given the roots of a quadratic equation, the equation is:

$$x^2 - (\text{sum of roots})\, x + \text{product of roots} = 0.$$

- If coefficients a, b and c are rational then the roots occur in conjugate pairs $d \pm \sqrt{e}$.

(d) **Useful Identities**

- $\alpha^2 + \beta^2 \equiv (\alpha + \beta)^2 - 2\alpha\beta \equiv (\alpha - \beta)^2 + 2\alpha\beta$.
- $(\alpha - \beta)^2 \equiv (\alpha + \beta)^2 - 4\alpha\beta$.
- $\alpha^3 + \beta^3 \equiv (\alpha + \beta)(\alpha^2 - \alpha\beta + \beta^2) \equiv (\alpha + \beta)\,[(\alpha + \beta)^2 - 3\alpha\beta]$.
- $\alpha^3 - \beta^3 \equiv (\alpha - \beta)(\alpha^2 + \alpha\beta + \beta^2) \equiv (\alpha - \beta)\,[(\alpha - \beta)^2 + 3\alpha\beta]$.

(e) **Completing the square**

- $x^2 + px$ requires $(p/2)^2$ to complete the square.
- $x^2 + px + (p/2)^2 = (x + p/2)^2$.

- If $f(x) = a(x + p)^2 + q$, and $a > 0$, then
  $f(x)$ has a least value of q, when $x = -p$ (diagram below, left).
- If $a < 0$, $f(x)$ has a greatest value of q, when $x = -p$ (diagram below, right).
- The graph of $y = f(x)$ has a line of symmetry at $x = -p$.

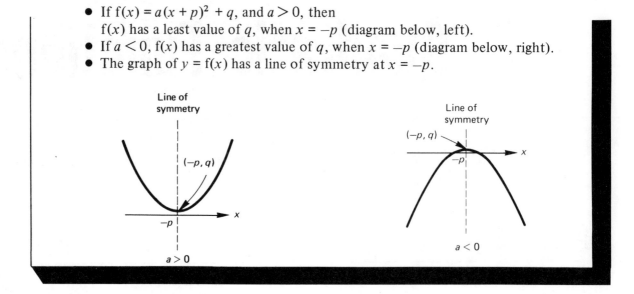

2.2 Worked Examples

2.1 The sum of the roots of a quadratic equation is 9 and the product of the roots is 4. The equation could be:

A, $9x^2 + 4x + 1 = 0$;   B, $4x^2 + 9x + 1 = 0$;
C, $x^2 + 9x - 4 = 0$;   D, $x^2 - 9x + 4 = 0$;
E, $x^2 - 4x + 9 = 0$.

- A quadratic equation can be written as
  $x^2 -$ (sum of roots) $x +$ product of roots $= 0$,
  i.e. $x^2 - 9x + 4 = 0$.   <u>Answer **D**</u>

2.2 Prove that the equation $x(x - 2p) = q(x - p)$ has real roots for all real values of p and q. If $q = -3$, find a non-zero value for p so that the roots are rational.

- $x(x - 2p) = q(x - p)$ $\Rightarrow$ $x^2 - 2px = qx - pq$

                        $\Rightarrow$ $x^2 - x(2p + q) + pq = 0$.

        Discriminant $= (2p + q)^2 - 4pq$,

                   $= 4p^2 + 4pq + q^2 - 4pq$,

                   $= 4p^2 + q^2$.

For all real values of p and q, $4p^2 \geqslant 0$ and $q^2 \geqslant 0$.
Hence the discriminant $\geqslant 0$ for all real p and q,
i.e. the equation has real roots.
If $q = -3$ the discriminant $= 4p^2 + 9$.

For rational roots the discriminant must be a perfect square and p must be rational.

Try $4p^2 + 9 = 9$, $4p^2 + 9 = 16$, $4p^2 + 9 = 25$, etc.
$4p^2 + 9 = 25$ is the first of these expressions which will give a rational non-zero value of p.

$$4p^2 = 16, \qquad p = 2.$$

Hence the roots are rational when $p = 2$ (or -2).

2.3 A quadratic equation with rational coefficients and one root $2 + \sqrt{3}$ is:

A, $x^2 + 4x - 1 = 0$;   B, $x^2 - 4x + 1 = 0$;
C, $x^2 + x + 4 = 0$;   D, $x^2 - 4x - 1 = 0$;
E, $x^2 + 4x + 1 = 0$.

- **Roots of a quadratic equation with rational coefficients occur in conjugate pairs $a + \sqrt{b}$, $a - \sqrt{b}$ where a and b are rational numbers.**

Roots of required equation are $2 + \sqrt{3}$ and $2 - \sqrt{3}$.
Sum of roots: $(2 + \sqrt{3}) + (2 - \sqrt{3}) = 4$.
Product of roots: $(2 + \sqrt{3})(2 - \sqrt{3}) = 4 - 3 = 1$.
Required equation is $x^2 - 4x + 1 = 0$.   Answer **B**

2.4 By completing the square, or otherwise, find the range of the function f given by

$$f(x) = x^2 - 6x + 10, \qquad x \in R.$$

Sketch the graph of $f(x)$.

- $f(x) = x^2 - 6x + 10$.
  $x^2 - 6x$ requires $+ 9$ to make a perfect square.
  Hence $f(x) = (x^2 - 6x + 9) + 1 = (x - 3)^2 + 1$,
  i.e. $f(x)$ has a minimum value of 1 when $x = 3$.
  Range of $f(x)$ is $f(x) \geqslant 1$.
  Graph of $f(x)$ is a parabola with a line of symmetry at $x = 3$, and a minimum turning point at $(3, 1)$.

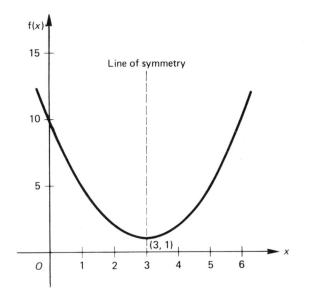

2.5 The quadratic function $f(x)$ takes the value 20 when $x = 1$ and $x = 5$ and takes the value 14 when $x = 2$. Obtain $f(x)$ in the form $ax^2 + bx + c$. Express $f(x)$ in the form $a(x - p)^2 + q$ and hence find the least possible value of $f(x)$.

Draw a rough sketch of the graph of $y = f(x)$ and state its relation to the graph of $y = ax^2$. The line through the origin and the point $(2, 14)$ meets the graph of $y = f(x)$ again at the point P. Find the coordinates of P.

- **If $f(1)$ and $f(5)$ are equal the line of symmetry is $x = \dfrac{1 + 5}{2}$, i.e. $x = 3$.**

11

$$f(x) = a(x - 3)^2 + q.$$

$$f(1) = 4a + q = 20. \tag{1}$$

$$f(2) = a + q = 14. \tag{2}$$

Equation 1 – equation 2:

$$3a = 6, \qquad a = 2, \qquad q = 12.$$

Therefore $f(x) = 2(x - 3)^2 + 12 = 2x^2 - 12x + 30$.

Since $f(x) = 2(x - 3)^2 + 12$, $f(x)$ has a least value of 12 when $x = 3$.

The graph $y = f(x)$ is a translation of $\begin{pmatrix} 3 \\ 12 \end{pmatrix}$ of the graph $y = 2x^2$.

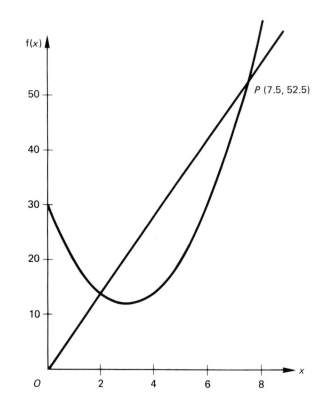

The line through the origin and $(2, 14)$ has a gradient of $\frac{14}{2}$, i.e. 7.

The equation of the line is $y = 7x$.

When the line cuts the curve $y = f(x)$ then $f(x) = 7x$,

i.e.
$$2x^2 - 12x + 30 = 7x$$
$$2x^2 - 19x + 30 = 0$$
$$(2x - 15)(x - 2) = 0$$

$$\Rightarrow \quad x = 2 \quad \text{or} \quad 7.5.$$

At P, $x = 7.5$, $\qquad y = 7(7.5) = 52.5$;

$$P(7.5, 52.5).$$

2.6 If x is real, find the set of possible values of $\dfrac{2x - 2}{x^2 + 3}$.

- Let $\dfrac{2x - 2}{x^2 + 3} = p$.

Then
$$2x - 2 = p(x^2 + 3)$$
$$\Rightarrow \quad px^2 - 2x + 3p + 2 = 0.$$

This is a quadratic equation in x and it is given that x is real. Therefore the equation must have real roots

i.e.
$$b^2 - 4ac \geqslant 0$$
$$\Rightarrow \quad (-2)^2 - 4p\,(3p + 2) \geqslant 0,$$
$$4 - 12p^2 - 8p \geqslant 0.$$
$$0 \geqslant 12p^2 + 8p - 4,$$
$$0 \geqslant 4\,(3p^2 + 2p - 1),$$
$$0 \geqslant 4\,(3p - 1)(p + 1).$$

If $f(p) = (3p - 1)\,(p + 1)$ then we require $f(p) \leqslant 0$.
The sketch of $f(p)$ shows that $f(p) \leqslant 0$ when $-1 \leqslant p \leqslant \frac{1}{3}$.

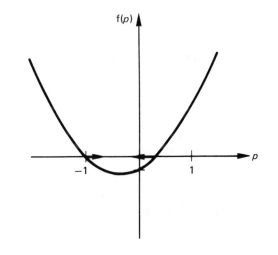

The set of possible values of $\dfrac{2x - 2}{x^2 + 3}$ is $-1 \leqslant \dfrac{2x - 2}{x^2 + 3} \leqslant \dfrac{1}{3}$.

2.7 Given that $px^2 + 2px - 5 < 0$ for all real values of x, determine the set of possible values of p.

- **Two conditions must be satisfied in this question:**
  (a) $px^2 + 2px - 5 = 0$ has no real solutions.
  (b) the graph of $px^2 + 2px - 5$ is entirely below the x-axis (i.e. has a maximum turning point).

  (a) $px^2 + 2px - 5 = 0$ has no real roots when
  $$b^2 - 4ac < 0.$$
  $$\Rightarrow \quad 4p^2 - 4(p)\,(-5) < 0$$
  $$4p^2 + 20p < 0$$
  $$4p(p + 5) < 0.$$

  If $f(p) = 4p(p + 5) < 0$ then $-5 < p < 0$.

  (b) $px^2 + 2px - 5$ has a maximum turning point when $p < 0$.
  Combining these two conditions gives $-5 < p < 0$.

2.8 Solve the equation $\sqrt{(4x-7)} + \sqrt{(2x-4)} = 1$ (where $\sqrt{}$ means positive root only).

- **In any equation with square root terms put one square root term on one side of the equation and all other terms on the other, then square both sides of the equation.**

If $\sqrt{(4x-7)} + \sqrt{(2x-4)} = 1$ then $\sqrt{(4x-7)} = 1 - \sqrt{(2x-4)}$.
Square both sides:

$$4x - 7 = (1 - \sqrt{(2x-4)})^2$$
$$= 1 + 2x - 4 - 2\sqrt{(2x-4)};$$

i.e.
$$2x - 4 = -2\sqrt{(2x-4)},$$
$$x - 2 = -\sqrt{(2x-4)}.$$

Square both sides:
$$x^2 - 4x + 4 = 2x - 4,$$
$$x^2 - 6x + 8 = 0,$$
$$(x-2)(x-4) = 0,$$
$$x = 2 \text{ or } 4.$$

Check both answers in the original equation:

$$x = 2: \text{l.h.s.} = \sqrt{1} + \sqrt{0} = 1 = \text{r.h.s.}$$
$$x = 4: \text{l.h.s.} = \sqrt{9} + \sqrt{4} = 5 \neq \text{r.h.s.}$$

Hence $x = 2$ is the only solution of the original equation.

Note: $\sqrt{(4x-7)} - \sqrt{(2x-4)} = 1$ would have led to the same values of x, $x = 2$ or 4.

In this case both $x = 2$ and $x = 4$ would be correct solutions – moral: *always* check.

2.9
(a) Solve the simultaneous equations $x^2 - y^2 = 24$ and $x - y = 2$.
(b) If $x^2 + 2xy + 3y^2 = 114$ and $x^2 - xy + y^2 = 19$ show that
$5x^2 - 8xy + 3y^2 = 0$.
Hence, or otherwise, find all possible pairs of values of x and y.

- (a) $x^2 - y^2 = 24$ and $x - y = 2$.

i.e.
$$(x-y)(x+y) = 24, \tag{1}$$
$$x - y = 2. \tag{2}$$

Substitute for $(x-y)$ into equation 1:

$$2(x+y) = 24,$$
$$x + y = 12. \tag{3}$$

Solve equations 2 and 3 simultaneously:

Adding, $\qquad 2x = 14, \qquad x = 7, y = 5.$

(b)
$$x^2 + 2xy + 3y^2 = 114 \qquad (1)$$
$$x^2 - xy + y^2 = 19. \qquad (2)$$

The required equation $5x^2 - 8xy + 3y^2 = 0$ has a constant term of zero. So eliminate the constant terms between equations 1 and 2.

Equation 2 x 6: $\qquad 6x^2 - 6xy + 6y^2 = 114.$ $\qquad (3)$
Equation 3 − equation 1:
$$5x^2 - 8xy + 3y^2 = 0.$$
Factorize:
$$(5x - 3y)(x - y) = 0, \qquad \text{i.e.} \qquad 5x = 3y \text{ or } x = y.$$
Substitute for y into equation 2:

(i) $y = \dfrac{5x}{3}$:

$$x^2 - x \left(\frac{5x}{3}\right) + \left(\frac{5x}{3}\right)^2 = 19,$$

$$x^2 - \frac{5x^2}{3} + \frac{25x^2}{9} = 19, \qquad 19x^2 = 171,$$

$$x^2 = 9, \qquad x = \pm 3, \, y = \pm 5.$$

(ii) $y = x$:

$$x^2 - x^2 + x^2 = 19, \qquad x = \pm\sqrt{19}, \qquad y = \pm\sqrt{19}.$$

Solutions are: $(3, 5); (-3, -5); (\sqrt{19}, \sqrt{19}); (-\sqrt{19}, -\sqrt{19}).$

2.10 If α and β are the roots of the equation $ax^2 + bx + c = 0$, form, without solving this equation, an equation whose roots are $\dfrac{\alpha^2}{\beta}$ and $\dfrac{\beta^2}{\alpha}$.

● For equation $ax^2 + bx + c = 0$,

$$\alpha + \beta = \frac{-b}{a}, \qquad \alpha\beta = \frac{c}{a}.$$

For required equation,

$$\text{sum of roots} = \frac{\alpha^2}{\beta} + \frac{\beta^2}{\alpha} = \frac{\alpha^3 + \beta^3}{\alpha\beta}$$

$$= \frac{(\alpha + \beta)(\alpha^2 - \alpha\beta + \beta^2)}{\alpha\beta}$$

$$= \frac{(\alpha + \beta)\,[(\alpha + \beta)^2 - 3\alpha\beta]}{\alpha\beta}$$

$$= \dfrac{-\dfrac{b}{a}\left(\dfrac{b^2}{a^2} - \dfrac{3c}{a}\right)}{\dfrac{c}{a}}.$$

Multiply numerator and denominator by a^3:

$$\text{sum of roots} = \dfrac{-b(b^2 - 3ca)}{ca^2} ;$$

$$\text{product of roots} = \left(\dfrac{\alpha^2}{\beta}\right)\left(\dfrac{\beta^2}{\alpha}\right) = \alpha\beta = \dfrac{c}{a} .$$

Required equation is

$$x^2 + \dfrac{b(b^2 - 3ac)x}{ca^2} + \dfrac{c}{a} = 0$$

or $\qquad\qquad ca^2 x^2 + b(b^2 - 3ac)x + c^2 a = 0.$

2.3 Exercises

2.1 If α and β are the roots of the equation $x^2 - 3x - 2 = 0$, find the quadratic equations whose roots are

(a) $\dfrac{2}{\alpha}, \dfrac{2}{\beta}$, (b) $\alpha^2\beta, \alpha\beta^2$.

2.2 The roots of the quadratic equation $x^2 - px + q = 0$ are α and β. Form, in terms of p and q, the quadratic equation whose roots are $\alpha^3 + p\alpha^2, \beta^3 + p\beta^2$.

2.3 If the equation $x^2 - qx + r = 0$ has roots $\alpha + 2, \beta - 1$, where α, β are the real roots of the equation $2x^2 - bx + c = 0$, and $\alpha \geqslant \beta$, find q and r in terms of b and c. In the case $\alpha = \beta$, show that $q^2 = 4r + 9$.

2.4 The roots of the equation $ax^2 + bx + c = 0$ are α and β. Derive the results

$\alpha + \beta = -\dfrac{b}{a}, \quad \alpha\beta = \dfrac{c}{a}.$

(You may assume the formula for the roots of a quadratic equation.)

2.5 The equations $ax^2 + bx + c = 0$ and $bx^2 + ax + c = 0$, where $a \neq b, c \neq 0$, have a common root. Prove that $a + b + c = 0$.

2.6 It is given that $f(x) = (x - 1)^2 - \mu(x + 3)(x + 2)$.
(a) Find the values of μ for which the equation $f(x) = 0$ has two equal roots.
(b) Show that when $\mu = 2$, $f(x)$ has a maximum value of 25.
(c) Given that the curve $y = f(x)$ has a turning point when $x = -\frac{3}{4}$, find the value of μ and sketch the curve for this value of μ.

2.7 Solve the equation

$$\sqrt{(3x - 2)} - \sqrt{(10 - x)} = 2$$

(where $\sqrt{\ }$ denotes positive root only).

2.8 Solve for a and b the simultaneous equations

$$a^2 + b^2 = \tfrac{13}{4}, \qquad ab = -\tfrac{3}{2}.$$

2.9 Find a quadratic equation of the form

$$ax^2 + bx + c = 0,$$

where a, b and c are integers, having $\dfrac{-4 + 3\sqrt{5}}{4}$ as one root.

2.4 Outline Solutions to Exercises

2.1 $\qquad\qquad\qquad \alpha + \beta = 3, \qquad \alpha\beta = -2.$

(a) $\dfrac{2}{\alpha} + \dfrac{2}{\beta} = \dfrac{2(\alpha + \beta)}{\alpha\beta} = -3; \qquad \left(\dfrac{2}{\alpha}\right)\left(\dfrac{2}{\beta}\right) = \dfrac{4}{\alpha\beta} = -2.$

Equation: $\qquad\qquad\qquad x^2 + 3x - 2 = 0.$

(b) $\alpha^2\beta + \alpha\beta^2 = \alpha\beta(\alpha + \beta) = -6; \qquad (\alpha^2\beta)(\alpha\beta^2) = \alpha^3\beta^3 = -8.$

Equation: $\qquad\qquad\qquad x^2 + 6x - 8 = 0.$

2.2 $\qquad\qquad \alpha + \beta = p, \qquad \alpha\beta = q.$

Sum: $\qquad \alpha^3 + p\alpha^2 + \beta^3 + p\beta^2 = (\alpha^3 + \beta^3) + p(\alpha^2 + \beta^2)$
$$= p(p^2 - 3q) + p(p^2 - 2q)$$
$$= 2p^3 - 5pq.$$

Product: $\qquad (\alpha^3 + p\alpha^2)(\beta^3 + p\beta^2) = \alpha^3\beta^3 + p\alpha^2\beta^2(\beta + \alpha) + p^2\alpha^2\beta^2$
$$= q^3 + 2p^2q^2.$$

Equation: $x^2 + (5pq - 2p^3)x + q^3 + 2p^2q^2 = 0.$

2.3 $\qquad\qquad 2x^2 - bx + c = 0: \qquad \alpha + \beta = \dfrac{b}{2}, \alpha\beta = \dfrac{c}{2}.$

$$x^2 - qx + r = 0$$
$$\Rightarrow \quad (\alpha + 2) + (\beta - 1) = q$$
$$\Rightarrow \quad q = \dfrac{b}{2} + 1.$$
$$(\alpha + 2)(\beta - 1) = r$$
$$\Rightarrow \alpha\beta - (\alpha + \beta) + 3\beta - 2 = r.$$

$\beta < \alpha \Rightarrow \beta = \dfrac{b - \sqrt{(b^2 - 8c)}}{4} \Rightarrow r = \dfrac{c}{2} + \dfrac{b}{4} - \dfrac{3\sqrt{(b^2 - 8c)}}{4} - 2.$

If $\alpha = \beta$, then $b^2 = 8c \Rightarrow r = \dfrac{c}{2} + \dfrac{b}{4} - 2 = \dfrac{b^2}{16} + \dfrac{b}{4} - 2$

$$= \dfrac{1}{4}\left(\dfrac{b}{2} + 1\right)^2 - \dfrac{9}{4}$$

$$\Rightarrow r = \frac{1}{4}q^2 - \frac{9}{4},$$

i.e.
$$q^2 = 4r + 9.$$

2.4 $ax^2 + bx + c = 0$: $\alpha = \dfrac{-b + \sqrt{(b^2 - 4ac)}}{2a}$, $\beta = \dfrac{-b - \sqrt{(b^2 - 4ac)}}{2a}$.

$$\alpha + \beta = \frac{-b + \sqrt{(b^2 - 4ac)} - b - \sqrt{(b^2 - 4ac)}}{2a} = \frac{-b}{a};$$

$$\alpha\beta = \left(\frac{b}{2a}\right)^2 - \frac{(b^2 - 4ac)}{(2a)^2} = \frac{4ac}{4a^2} = \frac{c}{a}.$$

2.5 Let α be the common root.
$$a\alpha^2 + b\alpha + c = 0,$$
$$b\alpha^2 + a\alpha + c = 0.$$
Subtract: $\alpha^2(b - a) + \alpha(a - b) = 0$.
$a \neq b$, so $\alpha = 0$, or $\alpha = +1$.
Since $c \neq 0$, $\alpha \neq 0$. Therefore $\alpha = 1$, giving $a + b + c = 0$.

2.6 $f(x) = (x - 1)^2 - \mu(x + 3)(x + 2)$
$\qquad\quad = x^2(1 - \mu) - x(2 + 5\mu) + 1 - 6\mu$.
(a) $f(x) = 0$ has equal roots.
$$(2 + 5\mu)^2 - 4(1 - \mu)(1 - 6\mu) = 0$$
$$\Rightarrow \mu^2 + 48\mu = 0, \mu = 0 \text{ or } -48.$$
(b) $\mu = 2$, $f(x) = -x^2 - 12x - 11$
$\qquad\qquad = -(x + 6)^2 + 25$.
$\quad$ $f(x)$ has a maximum value of 25 when $x = -6$.

(c) $f(x)$ has a turning point at $x = \dfrac{(2 + 5\mu)}{2(1 - \mu)}$.

$\quad$ Given $x = -\tfrac{3}{4} \Rightarrow \mu = -1$,
$$f(x) = 2x^2 + 3x + 7 = 2(x + \tfrac{3}{4})^2 + \tfrac{47}{8}.$$

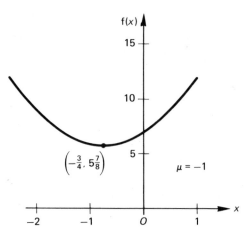

2.7 $\sqrt{(3x - 2)} = 2 + \sqrt{(10 - x)}$.
Squaring both sides gives $x - 4 = \sqrt{(10 - x)}$.
Squaring again yields $x^2 - 7x + 6 = 0$, $x = 6$ or 1.
Check When $x = 6$, $\sqrt{16} = 2 + \sqrt{4}$. When $x = 1$, $\sqrt{1} \neq 2 + \sqrt{9}$.
Therefore $x = 6$.

2.8 $a^2 + 2ab + b^2 = \frac{1}{4}$, so $a + b = \pm\frac{1}{2}$.

Also, $a^2 - 2ab + b^2 = \frac{25}{4}$, so $a - b = \pm\frac{5}{2}$.

$$a + b = \tfrac{1}{2}, a - b = \tfrac{5}{2} \qquad \Rightarrow \quad a = \tfrac{3}{2}, b = -1.$$

$$a + b = -\tfrac{1}{2}, a - b = \tfrac{5}{2} \qquad \Rightarrow \quad a = 1, b = -\tfrac{3}{2}.$$

$$a + b = \tfrac{1}{2}, a - b = -\tfrac{5}{2} \qquad \Rightarrow \quad a = -1, b = \tfrac{3}{2}.$$

$$a + b = -\tfrac{1}{2}, a - b = -\tfrac{5}{2} \quad \Rightarrow \quad a = -\tfrac{3}{2}, b = 1.$$

2.9 Roots are conjugate, $\dfrac{-4 + 3\sqrt{5}}{4}$ and $\dfrac{-4 - 3\sqrt{5}}{4}$.

Sum $= -2$, product $= -\frac{29}{16}$.

Equation: $x^2 + 2x - \frac{29}{16} = 0 \qquad \Rightarrow \qquad 16x^2 + 32x - 29 = 0.$

3 Progressions

Arithmetic and geometric progressions and sums to n terms. Sum to infinity of a geometric series. Use of Σ notation.

3.1 Fact Sheet

(a) Arithmetic Progression

- First term a, common difference d.
- nth term $= a + (n - 1)d$.
- Sum of n terms $= n(2a + (n - 1)d)/2$
  $\qquad\qquad = n$ (first term + last term)$/2$.
- Three terms p, q and r are in arithmetic progression if $p + r = 2q$.
- The arithmetic mean of p and r is $(p + r)/2$.

(b) Geometric Progression

- First term a, common ratio r.
- nth term $= ar^{n-1}$.

- Sum of n terms $S_n = \dfrac{a(1 - r^n)}{1 - r} = \dfrac{a(r^n - 1)}{r - 1}$.

- Sum to infinity exists if $|r| < 1$ or $-1 < r < 1$:

$$S_\infty = \frac{a}{1 - r}.$$

- Three terms x, y and z form a geometric progression if $xz = y^2$.
- The geometric mean of x and z is $\sqrt{(xz)}$.

(c) Σ Notation

- $\displaystyle\sum_{r=1}^{n} r^\alpha$ means the sum of all terms of the form r^α from 1^α to n^α, where n is a positive integer.

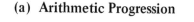

- $\displaystyle\sum_{r=1}^{n} ar^2 = a\sum_{r=1}^{n} r^2$, $\qquad \displaystyle\sum_{r=1}^{n}(ar^2 + br + c) = a\sum_{r=1}^{n} r^2 + b\sum_{r=1}^{n} r + c\sum_{r=1}^{n} 1$.

- $\displaystyle\sum_{r=1}^{n} 1 = n$.

(d) Useful Results

- $\sum\limits_{r=1}^{n} r = n(n+1)/2.$

- $\sum\limits_{r=1}^{n} r^2 = n(n+1)(2n+1)/6.$

- $\sum\limits_{r=1}^{n} r^3 = n^2(n+1)^2/4.$

Note: The letter r is often replaced by i.

- Sum of an arithmetic series

$$\sum_{i=1}^{n}[a+(i-1)d] = an + d\sum_{i=1}^{n}(i-1).$$

- Sum of a geometric series

$$\sum_{i=1}^{n} ar^{i-1} = a\sum_{i=1}^{n} r^{i-1}.$$

- Change of limits

$$\sum_{r=1}^{n}(r+a) = \sum_{r=a+1}^{a+n} r.$$

$$\sum_{r=m}^{n} f(r) = \sum_{r=1}^{n} f(r) - \sum_{r=1}^{m-1} f(r).$$

3.2 Worked Examples

3.1 The sum to infinity of the geometric series $a + a^2 + a^3 + \ldots\ (a \neq 0)$ is $4a$. The common ratio is

A, $\frac{4}{3}$;   B, $\frac{2}{3}$;   C, $\frac{3}{4}$;   D, $\frac{1}{4}$;   E, $\frac{3}{2}$.

- Sum to infinity $= \dfrac{a}{1-r}$.

But $r = a \quad \Rightarrow \quad 4a = \dfrac{a}{1-a} \quad \Rightarrow \quad 4a - 4a^2 = a$

$\Rightarrow \quad 4a^2 - 3a = 0 \quad \Rightarrow \quad a(4a-3) = 0$

$\Rightarrow \quad a = 0$ or $\frac{3}{4}$.

But $a \neq 0 \quad \Rightarrow \quad a = \frac{3}{4}$.   <u>Answer C</u>

3.2 An arithmetic progression has first term a and common difference d. Its fifth term is 59 and the sum of its first 30 terms is four times the sum of its first 10 terms. Find the values of a and d.

- Fifth term of an arithmetic progression: $a + 4d = 59$.   (1)

  Sum of first 30 terms $= 30(2a + 29d)/2 = 30a + 435d$.

  Sum of first 10 terms $= 10(2a + 9d)/2 = 10a + 45d$.

  Hence $30a + 435d = 4(10a + 45d)$.

  $\therefore 10a = 255d \qquad \Rightarrow \qquad 2a = 51d$,   (2)

  From equations 1 and 2, $a = 51$ and $d = 2$.

3.3 The first three terms of a geometric progression are $6, \frac{2}{3}, \frac{2}{27}$.

State the common ratio and show that the sum to infinity is 6.75.

Find the least number of terms required to make the sum exceed 6.7499.

- The common ratio is $\dfrac{\frac{2}{3}}{6}$,   i.e. $\frac{1}{9}$ $\Rightarrow$ $r = \frac{1}{9}$.

  Sum to infinity $S_\infty = \dfrac{a}{1-r} = \dfrac{6}{1-\frac{1}{9}} = 6.75$.

  Sum of n terms $S_n = \dfrac{a}{1-r}(1 - r^n) = S_\infty[1 - (\tfrac{1}{9})^n]$.

  When $S_n > 6.7499$, $6.75[1 - (\tfrac{1}{9})^n] > 6.7499$,

                         $6.75 - 6.75(\tfrac{1}{9})^n > 6.7499$,

                         $0.0001 > 6.75(\tfrac{1}{9})^n$.

  Multiply by $10^4 \times 9^n$: $9^n > 6.75 \times 10^4$.

  Take logs: $n \log 9 > \log(6.75 \times 10^4)$,

  $$n > \frac{4.8293}{0.9542},$$

  $$n > 5.0611.$$

Hence the least number of terms required is 6.

3.4

(a) An arithmetic progression is such that the sum of the first n terms is $2n^2$ for all positive integral values of n. Find, by substituting two values of n or otherwise, the first term and the common difference.

(b) The first term of a geometric progression is 20 and the sum to infinity is 40. Find the common ratio and the sum of the first 10 terms.

- (a) $S_n = 2n^2$.

  $S_n = u_1 + u_2 + u_3 + \ldots$ where u_1 etc. are the terms of the series

  $S_1 = u_1, S_2 = u_1 + u_2 \quad \Rightarrow \quad u_2 = S_2 - S_1$.

  $S_1 = 2(1)^2 = 2$, i.e. first term $= 2$.

  $S_2 = 2(2)^2 = 8$, i.e. second term $= 6$.

  Common difference $= u_2 - u_1 = 4$.

  (b) $S_\infty = \dfrac{a}{1-r}$,   i.e. $40 = \dfrac{20}{1-r}$ $\Rightarrow$ $r = \frac{1}{2}$,

  i.e. common ratio $= \frac{1}{2}$.

  $S_{10} = 40[1 - (\tfrac{1}{2})^{10}] = \frac{5115}{128}$.

  Hence the sum of the first 10 terms $= \frac{5115}{128}$.

3.5 Given that the sum of the first 2 terms of a geometric progression is 16 and the sum to infinity is 18, find the possible values of the common ratio.

- If the first term is a and the common ratio is r then

$$S_2 = a + ar = a(1+r) = 16, \qquad (1)$$

$$S_\infty = \frac{a}{1-r} = 18. \qquad (2)$$

From equation 2, $a = 18(1-r)$.
Substituting for a in equation 1,
$$18(1-r)(1+r) = 16 \quad \Rightarrow \quad 9(1-r^2) = 8,$$
$$9r^2 = 1 \quad \Rightarrow \quad r = \pm\tfrac{1}{3}.$$
Hence the common ratio is $\pm\tfrac{1}{3}$.

3.6 By using an infinite geometric progression show that $0.65\dot{2}\dot{4}$, i.e. $0.652\,424\,24\ldots$ is equal to $\frac{2153}{3300}$.

- $0.652\,424\,24\ldots = 0.65 + 0.0024 + 0.000\,024 + \ldots$
$$= 0.65 + 24(10^{-4} + 10^{-6} + 10^{-8} + \ldots).$$
$10^{-4} + 10^{-6} + \ldots$ is a GP with $a = 10^{-4}$ and $r = 10^{-2}$.

$$\text{Sum of GP} = \frac{10^{-4}}{(1-10^{-2})} = \tfrac{1}{9900}$$

Hence $0.65\dot{2}\dot{4} = \tfrac{65}{100} + \tfrac{24}{9900} = \tfrac{2153}{3300}$.

3.7 Find, in terms of n, the value of $\sum_{r=1}^{n}(r^2 + 2^r)$.

- $$\sum_{r=1}^{n}(r^2 + 2^r) = \sum_{r=1}^{n} r^2 + \sum_{r=1}^{n} 2^r$$

$$= \frac{n(n+1)(2n+1)}{6} + (2 + 2^2 + 2^3 + \ldots 2^n)$$

$$= \frac{n(n+1)(2n+1)}{6} + \frac{2(2^n - 1)}{(2-1)}$$

$$= \frac{n(n+1)(2n+1)}{6} + 2(2^n - 1).$$

3.8
(a) The sum of the first n terms of a series is $1 - (\tfrac{1}{4})^n$. Obtain the values of the first three terms of this series. What is the sum to infinity of this series?
(b) The rth term of a series is $2^r + 3r - 2$. Find a formula for the sum of the first n terms.

- (a) If $S_n = u_1 + u_2 + \ldots + u_n$,

then $S_1 = 1 - \tfrac{1}{4}$; $u_1 = \tfrac{3}{4}$.

$S_2 = 1 - \tfrac{1}{16} = \tfrac{15}{16}$; $u_2 = S_2 - S_1$; $u_2 = \tfrac{3}{16}$.

$S_3 = 1 - \tfrac{1}{64} = \tfrac{63}{64}$; $u_3 = S_3 - S_2$; $u_3 = \tfrac{3}{64}$.

Hence the first three terms of this series are

$$\tfrac{3}{4}, \tfrac{3}{16} \text{ and } \tfrac{3}{64}.$$

This is a GP with first term $\tfrac{3}{4}$ and common ratio $\tfrac{1}{4}$.

$$S_\infty = (\tfrac{3}{4})/(1 - \tfrac{1}{4}) = 1.$$

(This could also have been deduced from the limit of S_n as $n \to \infty$.)

(b) rth term $u_r = 2^r + 3r - 2$.

$$u_1 = 2^1 + 3(1) - 2, \qquad u_2 = 2^2 + 3(2) - 2, \qquad \text{etc.}$$

$$S_n = \sum_{r=1}^{n}(2^r + 3r - 2) = \sum_{r=1}^{n} 2^r + 3\sum_{r=1}^{n} r - 2\sum_{r=1}^{n} 1.$$

$$= \frac{2(2^n - 1)}{(2 - 1)} + \frac{3n(n + 1)}{2} - 2n$$

$$= 2^{n+1} - 2 + \frac{3n^2}{2} - \frac{n}{2}$$

$$= 2^{n+1} + \frac{(3n^2 - n - 4)}{2}.$$

$$\text{Sum of first } n \text{ terms} = 2^{n+1} + \frac{(3n - 4)(n + 1)}{2}.$$

3.3 Exercises

3.1 The first term of an arithmetic progression is 17, and the sum of the first 16 terms is –16. Find the sixteenth term and the common difference of the progression.

3.2 Starting from first principles, prove that the sum of the first n terms of a geometric progression whose first term is a and whose common ratio is r (where $r \neq 1$) is

$$\frac{a(1 - r^n)}{(1 - r)}.$$

The first and second terms of an infinite geometric progression are 16 and 8 respectively. Show that the sum of all the terms after the nth term is 2^{5-n}.

3.3 Write out in full the terms of the series

$$\sum_{i=0}^{3}(-1)^i\left(\frac{i + 1}{3i - 1}\right) x^i.$$

3.4 By using an infinite geometric progression, show that $0.42\dot{3}\dot{4}$, i.e. $0.423\,434\,34\ldots$, is equal to $\frac{1048}{2475}$.

3.5 Evaluate $\sum_{r=1}^{n}(3r + 2)^2$.

3.6 The sum of the first twenty terms of an arithmetic series is 45 and the sum of the first forty terms is 490. Find the first term and the common difference.

3.7 If p, q and r are three successive terms of a geometric progression show that $\log p$, $\log q$ and $\log r$ are three successive terms of an arithmetic progression. (p, q, and r are > 0.)

3.8 The sum of the first n terms of a series is given by $S_n = 16n - n^2$. Show that the terms are in arithmetic progression and find the tenth term.

3.4 Outline Solutions to Exercises

3.1 $a = 17$; $-16 = \dfrac{16(17 + 16\text{th term})}{2}$,

$$-2 = 17 + 16\text{th term} \quad \Rightarrow \quad 16\text{th term} = -19.$$

$$-19 = 17 + 15d \quad \Rightarrow \quad d = -\tfrac{12}{5}.$$

3.2 $S_n = a + ar + ar^2 + \ldots + ar^{n-1}$,

$rS_n = \qquad ar + ar^2 + \ldots + ar^{n-1} + ar^n$.

Subtracting, $S_n(1 - r) = a - ar^n = a(1 - r^n)$

$$S_n = \frac{a(1 - r^n)}{1 - r}.$$

$a = 16$, $r = \tfrac{1}{2}$, $\Rightarrow$ $S_n = \dfrac{16[1 - (\tfrac{1}{2})^n]}{1 - \tfrac{1}{2}} = 32 - 32\,(\tfrac{1}{2})^n$.

$$S_\infty = \frac{16}{1 - \tfrac{1}{2}} = 32.$$

Sum of all the terms after nth term $= S_\infty - S_n = 32(\tfrac{1}{2})^n$,

$$= 2^{5-n}.$$

3.3 $\displaystyle\sum_{i=0}^{3} (-1)^i \left(\frac{i+1}{3i-1}\right) x^i = (-1)^0 \left(\frac{1}{-1}\right) x^0 + (-1)^1 \left(\frac{2}{2}\right) x^1 + (-1)^2 \left(\frac{3}{5}\right) x^2$

$$+ (-1)^3 \left(\frac{4}{8}\right) x^3$$

$$= -1 - x + \tfrac{3}{5}x^2 - \tfrac{1}{2}x^3.$$

3.4 $0.42\dot{3}\dot{4} = \tfrac{42}{100} + 34 \times 10^{-4} + 34 \times 10^{-6} + \ldots$

$$= \tfrac{42}{100} + \frac{34 \times 10^{-4}}{1 - 10^{-2}}$$

$$= \tfrac{42}{100} + \tfrac{34}{9900} = \tfrac{1048}{2475}.$$

3.5 $\displaystyle\sum_{r=1}^{n}(3r+2)^2 = \sum_{r=1}^{n}(9r^2+12r+4) = 9\sum_{r=1}^{n}r^2 + 12\sum_{r=1}^{n}r + 4n$

$$= 9n(n+1)(2n+1)/6 + 6n(n+1) + 4n$$

$$= n[3(n+1)(2n+1) + 12n + 12 + 8]/2$$

$$= n(6n^2 + 21n + 23)/2.$$

3.6 $\qquad\qquad 45 = 10(2a + 19d) \quad\Rightarrow\quad 9 = 4a + 38d,$ $\qquad\qquad$ (1)

$\qquad\qquad 490 = 20(2a + 39d) \quad\Rightarrow\quad 49 = 4a + 78d.$ $\qquad\qquad$ (2)

Equation 2 – equation 1: $\qquad\qquad\qquad\qquad 40 = 40d;$

$\therefore d = 1, a = -\frac{29}{4}.$

3.7 $pr = q^2$ for three terms in GP.

Take logs: $\log p + \log r = 2 \log q$.

But x, y and z are in AP if $x + z = 2y$.

Hence $\log p$, $\log q$ and $\log r$ are successive terms of an AP.

3.8 $S_n = 16n - n^2.$ $\quad S_{n-1} = 16(n-1) - (n-1)^2.$

$u_n = [16n - n^2] - [16(n-1) - (n-1)^2] = -2n + 17,$

$u_{n-1} = -2(n-1) + 17 = -2n + 19;$

$u_n - u_{n-1} = -2 \quad\Rightarrow\quad$ common difference is -2, i.e. the terms are in AP.

$u_{10} = -20 + 17 = -3.$

4 Binomial Expansions

The use of the binomial expansion of $(1 + x)^n$, when

(a) n is a positive integer,

(b) n is rational and $|x| < 1$.

4.1 Fact Sheet

(a) If n is a Positive Integer:

$$(1 + x)^n = 1 + nx + \frac{n(n-1)}{(1)(2)} x^2 + \frac{n(n-1)(n-2)}{(1)(2)(3)} x^3 + \ldots + nx^{n-1} + x^n$$

$$= 1 + {}_nC_1 x + {}_nC_2 x^2 + {}_nC_3 x^3 + \ldots + {}_nC_r x^r + \ldots + {}_nC_{n-1} x^{n-1} + x^n$$

$$= 1 + \binom{n}{1} x + \binom{n}{2} x^2 + \binom{n}{3} x^3 + \ldots + \binom{n}{r} x^r + \ldots + \binom{n}{n-1} x^{n-1} + x^n$$

where $\binom{n}{r} = {}_nC_r = \dfrac{n!}{r!\,(n-r)!} = \dfrac{n(n-1)(n-2)\ldots(n-r+1)}{r(r-1)(r-2)\ldots(2)(1)}$.

(i) The expansion terminates automatically after $n + 1$ terms.

(ii) The binomial coefficients may be speedily determined by Pascal's triangle:

$$
\begin{array}{ccccccccccc}
 & & & & & 1 & & & & & \\
 & & & & 1 & & 1 & & & & \\
 & & & 1 & & 2 & & 1 & & & \\
 & & 1 & & 3 & & 3 & & 1 & & \\
 & 1 & & 4 & & 6 & & 4 & & 1 & \\
1 & & 5 & & 10 & & 10 & & 5 & & 1
\end{array}
$$

(iii) $(a + b)^n = a^n + na^{n-1}b + \dfrac{n(n-1)}{(1)(2)} a^{n-2}b^2 + \ldots + nab^{n-1} + b^n$

$$= a^n + \binom{n}{1} a^{n-1}b + \binom{n}{2} a^{n-2}b^2 + \ldots + \binom{n}{n-1} ab^{n-1} + b^n$$

$$= \sum_{r=0}^{n} \binom{n}{r} a^{n-r}\, b^r.$$

(b) If n is NOT a Positive Integer

$$(1 + x)^n = 1 + nx + \frac{n(n-1)}{(1)(2)} x^2 + \frac{n(n-1)(n-2)}{(1)(2)(3)} x^3 + \ldots \qquad \text{if } |x| < 1.$$

This series is an infinite series, which can only be used when $|x| < 1$, i.e. $-1 < x < +1$.

(c) **Particular Series**:

$$(1 + x)^{-1} = 1 - x + x^2 - x^3 + \ldots$$
$$(1 - x)^{-1} = 1 + x + x^2 + x^3 + \ldots$$
$$(1 + x)^{-2} = 1 - 2x + 3x^2 - 4x^3 + \ldots$$
$$(1 - x)^{-2} = 1 + 2x + 3x^2 + 4x^3 + \ldots$$

4.2 Worked Examples

4.1 Find, and simplify the middle term in the expansion, in ascending powers of x, of $(3 - 5x)^8$.

● **The middle term of $(a + bx)^n$ is the term containing $x^{n/2}$ (n even).**

The middle term is the 5th term and contains $(-5x)^4$.

Middle term is $\binom{8}{4} (3)^4 (-5x)^4$.

i.e. $\dfrac{8 \times 7 \times 6 \times 5}{1 \times 2 \times 3 \times 4} \times (3)^4 \times (-5)^4 \times x^4$

i.e. $70 \times 81 \times 625x^4$,

i.e. $3\,543\,750x^4$.

4.2 Obtain the term independent of x in the expansion of $(3x - 1/x^2)^{15}$, leaving the answer in terms of factorials.

● Express $\left(3x - \dfrac{1}{x^2}\right)^{15}$ as $(3x)^{15} \left(1 - \dfrac{1}{3x^3}\right)^{15}$

$$= (3)^{15} (x)^{15} (1 - \tfrac{1}{3} x^{-3})^{15}.$$

The expansion of $\left(1 - \dfrac{1}{3}x^{-3}\right)^{15} = \displaystyle\sum_{r=0}^{15} \binom{15}{r}\left(-\dfrac{1}{3}x^{-3}\right)^r$.

The required term of this series must contain $(x)^{-15}$,
i.e. $(x^{-3})^r = x^{-15}$, $\therefore r = 5$.

Therefore the term independent of x is $3^{15} \, x^{15} \, \binom{15}{5} \left(-\dfrac{1}{3}\right)^5 x^{-15}$.

i.e. $3^{15} \, \dfrac{15!}{10!\,5!} \left(-\dfrac{1}{3}\right)^5$,

i.e. $\dfrac{-3^{10}\,15!}{10!\,5!}$.

4.3 Use the binomial expansion to find a quadratic approximation for

$$\frac{1}{(1 + 2x)^{1/3}} - \frac{1}{(9 - 4x)^{3/2}}$$

where x is small enough for terms in x^3 and higher powers to be negligible.

• $\dfrac{1}{(1+2x)^{1/3}} = (1+2x)^{-1/3}$

$$= 1 + (-\tfrac{1}{3})(2x) + \dfrac{(-\tfrac{1}{3})(-\tfrac{4}{3})}{(1)(2)}(2x)^2 + \dots \text{ higher powers of } x.$$

$$= 1 - \tfrac{2}{3}x + \tfrac{8}{9}x^2 + \dots \qquad \text{if } |2x| < 1, \text{ i.e. } |x| < \tfrac{1}{2}.$$

When the power is fractional or negative, remember to make the first term in the bracket equal to 1.

$$\dfrac{1}{(9-4x)^{3/2}} = \dfrac{1}{9^{3/2}(1-\tfrac{4}{9}x)^{3/2}} = \dfrac{(1-\tfrac{4}{9}x)^{-3/2}}{27}.$$

$$\tfrac{1}{27}(1-\tfrac{4}{9}x)^{-3/2} = \tfrac{1}{27}\left[1 + (-\tfrac{3}{2})(-\tfrac{4}{9}x) + \dfrac{(-\tfrac{3}{2})(-\tfrac{3}{2}-1)(-\tfrac{4}{9}x)^2}{(1)(2)} + \dots\right]$$

$$= \tfrac{1}{27}(1 + \tfrac{2}{3}x + \tfrac{10}{27}x^2 + \dots) \qquad \text{if } |\tfrac{4}{9}x| < 1, \text{ i.e. } |x| < \tfrac{9}{4}.$$

Hence

$$\dfrac{1}{(1+2x)^{1/3}} - \dfrac{1}{(9-4x)^{3/2}} = (1 - \tfrac{2}{3}x + \tfrac{8}{9}x^2 + \dots) - (\tfrac{1}{27} + \tfrac{2}{81}x + \tfrac{10}{729}x^2 + \dots)$$

$$\approx \tfrac{26}{27} - \tfrac{56}{81}x + \tfrac{638}{729}x^2.$$

Taking the smaller range of x, series is valid if $|x| < \tfrac{1}{2}$.

4.4 Given that $f(x) = \dfrac{x}{(1-x)(1-2x)^2}$, show that $f(x)$ can be expressed as

$$\dfrac{1}{(1-x)} - \dfrac{2}{(1-2x)} + \dfrac{1}{(1-2x)^2}.$$

Hence, or otherwise, find the first three terms in the expansion of $f(x)$ in ascending powers of x. State the range of values of x for which the expansion is valid.

• Let $g(x) = \dfrac{1}{(1-x)} - \dfrac{2}{(1-2x)} + \dfrac{1}{(1-2x)^2}$

$$= \dfrac{(1-2x)^2 - 2(1-x)(1-2x) + (1-x)}{(1-x)(1-2x)^2}$$

$$= \dfrac{1 - 4x + 4x^2 - 2 + 6x - 4x^2 + 1 - x}{(1-x)(1-2x)^2}$$

$$= \dfrac{x}{(1-x)(1-2x)^2}.$$

Hence $g(x) \equiv f(x)$.

So $\qquad f(x) \equiv (1-x)^{-1} - 2(1-2x)^{-1} + (1-2x)^{-2}.$

By the binomial expansion,

$$(1-x)^{-1} = 1 + (-1)(-x) + \frac{(-1)(-2)}{(1)(2)}(-x)^2 + \dots$$

$$= 1 + x + x^2 + x^3 + \dots \text{ valid for } -1 < x < 1.$$

$(1 - 2x)^{-1}$ **may be written down directly if x is replaced by $2x$ in the expansion of $(1-x)^{-1}$.**

$$2(1-2x)^{-1} = 2[1 + (2x) + (2x)^2 + (2x)^3 + \dots]$$
$$= 2 + 4x + 8x^2 + 16x^3 + \dots$$

valid for $-1 < 2x < 1$, that is $-\frac{1}{2} < x < \frac{1}{2}$.

$$(1-2x)^{-2} = 1 + (-2)(-2x) + \frac{(-2)(-3)}{(1)(2)}(-2x)^2 + \frac{(-2)(-3)(-4)}{(1)(2)(3)}(-2x)^3 + \dots$$

$$= 1 + 4x + 12x^2 + 32x^3 + \dots \qquad (\text{valid } -\tfrac{1}{2} < x < \tfrac{1}{2}).$$

Hence
$$f(x) \equiv (1 + x + x^2 + x^3 + \dots) - (2 + 4x + 8x^2 + 16x^3 + \dots)$$

$$+ (1 + 4x + 12x^2 + 32x^3 + \dots)$$

$$= x + 5x^2 + 17x^3 + \dots.$$

This is valid when $-1 < x < 1$ AND $-\frac{1}{2} < x < \frac{1}{2}$.
that is, when $-\frac{1}{2} < x < \frac{1}{2}$.

4.5 Find the expansion of $(1 - x + 2x^2)^{1/2}$ up to and including the term in x^4.

- $(1 - x + 2x^2)^{1/2}$ **must be expressed in the form $(1 + z)^n$, where $z = -x + 2x^2$.**

By the binomial expansion:

$$[1 + (-x + 2x^2)]^{1/2} = 1 + (\tfrac{1}{2})(-x + 2x^2) + \frac{(\tfrac{1}{2})(-\tfrac{1}{2})}{(1)(2)}(-x + 2x^2)^2$$

$$+ \frac{(\tfrac{1}{2})(-\tfrac{1}{2})(-\tfrac{3}{2})}{(1)(2)(3)}(-x + 2x^2)^3 + \frac{(\tfrac{1}{2})(-\tfrac{1}{2})(-\tfrac{3}{2})(-\tfrac{5}{2})}{(1)(2)(3)(4)}(-x + 2x^2)^4 + \dots$$

$$= 1 + \tfrac{1}{2}(-x + 2x^2) - \tfrac{1}{8}(-x + 2x^2)^2 + \tfrac{1}{16}(-x + 2x^2)^3 - \tfrac{5}{128}(-x + 2x^2)^4 + \dots$$

Now, $(-x + 2x^2)^2 = x^2 - 4x^3 + 4x^4$
$(-x + 2x^2)^3 = (-x)^3 + 3(-x)^2(2x^2) + \dots = -x^3 + 6x^4 + \dots$
$(-x + 2x^2)^4 = (-x)^4 + \dots = x^4 + \dots.$

Therefore

$(1 - x + 2x^2)^{1/2}$

$$= 1 - \tfrac{1}{2}x + x^2 - \tfrac{1}{8}(x^2 - 4x^3 + 4x^4) + \tfrac{1}{16}(-x^3 + 6x^4 + \dots) - \tfrac{5}{128}(x^4 + \dots) + \dots$$

$$= 1 - \tfrac{1}{2}x + (1 - \tfrac{1}{8})x^2 + (\tfrac{1}{2} - \tfrac{1}{16})x^3 + (-\tfrac{1}{2} + \tfrac{3}{8} - \tfrac{5}{128})x^4 + \dots$$

$$= 1 - \tfrac{1}{2}x + \tfrac{7}{8}x^2 + \tfrac{7}{16}x^3 - \tfrac{21}{128}x^4 + \dots.$$

4.6 Find the exact value of $(\sqrt{7} + 2)^4 + (\sqrt{7} - 2)^4$ and hence find the value of the integer n such that
$$n < (\sqrt{7} + 2)^4 < n + 1.$$

- $(\sqrt{7} + 2)^4 = (\sqrt{7})^4 + 4(\sqrt{7})^3(2) + 6(\sqrt{7})^2(4) + 4(\sqrt{7})(8) + 16.$

  $(\sqrt{7} - 2)^4 = (\sqrt{7})^4 - 4(\sqrt{7})^3(2) + 6(\sqrt{7})^2(4) - 4(\sqrt{7})(8) + 16.$

  So $(\sqrt{7} + 2)^4 + (\sqrt{7} - 2)^4 = 2[(\sqrt{7})^4 + 6(\sqrt{7})^2(4) + 16].$

  $(\sqrt{7})^2 = 7, (\sqrt{7})^4 = 49.$

  Hence $(\sqrt{7} + 2)^4 + (\sqrt{7} - 2)^4 = 2[49 + 168 + 16] = 466.$

  Since $2 < \sqrt{7} < 3$, subtracting 2 from each part gives

  $$0 < \sqrt{7} - 2 < 1 \Rightarrow 0 < (\sqrt{7} - 2)^4 < 1.$$

  Hence $(\sqrt{7} + 2)^4$ lies between 465 and 466, i.e. $n = 465$.

4.7 If $x + \dfrac{1}{x} = t$, express $x^3 + \dfrac{1}{x^3}$ and $x^5 + \dfrac{1}{x^5}$ in terms of t.

- Consider $\left(x + \dfrac{1}{x}\right)^3 = x^3 + 3(x)^2\left(\dfrac{1}{x}\right) + 3(x)\left(\dfrac{1}{x}\right)^2 + \left(\dfrac{1}{x}\right)^3$

  $$= x^3 + 3x + \dfrac{3}{x} + \dfrac{1}{x^3}.$$

  Since the binomial coefficients are symmetrical, these terms may be paired as

  $$\left(x^3 + \dfrac{1}{x^3}\right) + 3\left(x + \dfrac{1}{x}\right).$$

  Therefore $\left(x + \dfrac{1}{x}\right)^3 = t^3 = \left(x^3 + \dfrac{1}{x^3}\right) + 3t$

  $$\Rightarrow \quad x^3 + \dfrac{1}{x^3} = t^3 - 3t.$$

  Similarly,

  $$\left(x + \dfrac{1}{x}\right)^5 = x^5 + 5(x)^4\left(\dfrac{1}{x}\right) + 10(x)^3\left(\dfrac{1}{x}\right)^2 + 10(x)^2\left(\dfrac{1}{x}\right)^3 + 5(x)\left(\dfrac{1}{x}\right)^4 + \left(\dfrac{1}{x}\right)^5$$

  $$= x^5 + 5x^3 + 10x + \dfrac{10}{x} + \dfrac{5}{x^3} + \dfrac{1}{x^5}$$

  $$= \left(x^5 + \dfrac{1}{x^5}\right) + 5\left(x^3 + \dfrac{1}{x^3}\right) + 10\left(x + \dfrac{1}{x}\right).$$

  Therefore $t^5 = \left(x^5 + \dfrac{1}{x^5}\right) + 5(t^3 - 3t) + 10t$

  $$\Rightarrow \quad x^5 + \dfrac{1}{x^5} = t^5 - 5t^3 + 5t.$$

4.3 Exercises

4.1 Express $\dfrac{\sqrt{(1 + x)}}{1 - x}$ in ascending powers of x up to and including the term in x^3. By substituting $x = \frac{1}{4}$, show that $\sqrt{5} \approx \frac{4557}{2048}$.

4.2 Show that the exact value of $(7 + 3\sqrt{5})^4 + (7 - 3\sqrt{5})^4$ is an integer. Hence, find two consecutive integers n and $n + 1$ such that $n < (7 + 3\sqrt{5})^4 < n + 1$.

4.3
(a) Expand and simplify $\left(x + \dfrac{1}{3x}\right)^5 + \left(x - \dfrac{1}{3x}\right)^5$.

(b) Find the coefficient of x^3 in the expansion of $\left(x + \dfrac{1}{3x}\right)^4 \left(x - \dfrac{1}{3x}\right)^3$.

4.4 State the binomial expansion of $(x + y)^5$, giving the coefficients as integers. Given that $x + y = p$ and $xy = q$, express $x^5 + y^5$ in terms of p and q.

4.5 Find the values of the constants a and n so that the expansions of

$$\left(1 + 3x + \frac{3x^2}{2}\right)^4 \qquad \text{and} \qquad (1 + ax)^n$$

agree as far as the term in x^2. For these values of a and n, determine the difference in the coefficients of the term in x^3 in the expansions.

4.6
(a) Find the expansion of $(1 + 2x + 3x^2)^5$ in ascending powers of x up to and including the term in x^3.
(b) Calculate the coefficient of the term, independent of x, in the expansion of

$$\left(x^2 + \frac{1}{x}\right)^{12}.$$

4.7 Verify that $f(x) = \dfrac{9}{(1 - x)(1 + 2x)^2}$ may be expressed as

$$f(x) = \frac{1}{1 - x} + \frac{2}{1 + 2x} + \frac{6}{(1 + 2x)^2}.$$

Hence show that $f(x) = 9 - 27x + 81x^2$, when x is sufficiently small to allow x^3 and higher powers of x to be neglected.

4.8 Given that $|3x^3| < 1$, write down the first four terms in the binomial expansion of $(1 + 3x^3)^{1/3}$ in ascending powers of x. By putting $x = 0.2$, estimate $\sqrt[3]{2}$ to five decimal places.

4.9 Prove that, for $|x| < 1$, $\sqrt{\left(\dfrac{1 + x}{1 - x}\right)} \approx 1 + x + \dfrac{x^2}{2} + \dfrac{x^3}{2}$.

By suitable choice of a value for x, prove that

$$\sqrt{5} \approx \frac{1630}{729}.$$

4.10 The numbers $c_0, c_1, c_2, \ldots, c_n$ are the binomial coefficients such that

$$(1+x)^n = c_0 + c_1 x + c_2 x^2 + c_3 x^3 + \ldots + c_n x^n.$$

Prove that

(a) $c_0 + c_2 + c_4 + \ldots + c_{2n} = 2^{2n-1}$.

(b) $c_0 + 2c_1 + 3c_2 + \ldots + (n+1) c_n = 2^n + n 2^{n-1}$.

(c) $c_0 - \frac{1}{2} c_1 + \frac{1}{3} c_2 - \ldots + \frac{(-1)^n}{n+1} c_n = \frac{1}{n+1}$.

4.4 Outline Solutions to Exercises

4.1 $(1+x)^{1/2} = 1 + \frac{1}{2} x - \frac{1}{8} x^2 + \frac{1}{16} x^3 + \ldots$ for $|x| < 1$.

$(1-x)^{-1} = 1 + x + x^2 + x^3 + \ldots$ for $|x| < 1$.

Hence $\dfrac{\sqrt{(1+x)}}{(1-x)} = 1 + \frac{3}{2} x + \frac{11}{8} x^2 + \frac{23}{16} x^3 + \ldots$.

When $x = \frac{1}{4}$, $\dfrac{\sqrt{\frac{5}{4}}}{\frac{3}{4}} = 1 + \frac{3}{8} + \frac{11}{128} + \frac{23}{1024} + \ldots$

$= \frac{1519}{1024}.$

Hence $\sqrt{5} \approx \frac{3}{2} \left(\frac{1519}{1024} \right) = \frac{4557}{2048}$.

4.2 $(7 + 3\sqrt{5})^4 = 7^4 + 4(7)^3 (3\sqrt{5}) + 6(7)^2 (3\sqrt{5})^2 + 4(7)(3\sqrt{5})^3 + (3\sqrt{5})^4$

$(7 - 3\sqrt{5})^4 = 7^4 - 4(7)^3 (3\sqrt{5}) + 6(7)^2 (3\sqrt{5})^2 - 4(7)(3\sqrt{5})^3 + (3\sqrt{5})^4$.

Hence

$$(7 + 3\sqrt{5})^4 + (7 - 3\sqrt{5})^4 = 2[7^4 + (6)(49)(45) + (81)(25)] = 35\,312. \qquad (1)$$

Now $36 < 45 < 49 \;\Rightarrow\; 6 < 3\sqrt{5} < 7 \;\Rightarrow\; -1 < (3\sqrt{5} - 7) < 0$.

Multiply by -1 (inequalities reverse):

$$0 < (7 - 3\sqrt{5}) < 1 \;\Rightarrow\; 0 < (7 - 3\sqrt{5})^4 < 1.$$

Hence (from equation 1)

$$35\,311 < (7 + 3\sqrt{5})^4 < 35\,312 \;\Rightarrow\; n = 35\,311.$$

4.3

(a) $\left(x + \dfrac{1}{3x} \right)^5 = x^5 + \dfrac{5x^3}{3} + \dfrac{10x}{9} + \dfrac{10}{27x} + \dfrac{5}{81x^3} + \dfrac{1}{243x^5}$,

$\left(x - \dfrac{1}{3x} \right)^5 = x^5 - \dfrac{5x^3}{3} + \dfrac{10x}{9} - \dfrac{10}{27x} + \dfrac{5}{81x^3} - \dfrac{1}{243x^5}$.

$\Rightarrow \left(x + \dfrac{1}{3x} \right)^5 + \left(x - \dfrac{1}{3x} \right)^5 = 2x^5 + \dfrac{20x}{9} + \dfrac{10}{81x^3}$.

(b) $\left(x + \dfrac{1}{3x} \right)^4 = x^4 + \dfrac{4x^2}{3} + \dfrac{2}{3} + \dfrac{4}{27x^2} + \dfrac{1}{81x^4}$,

$\left(x - \dfrac{1}{3x} \right)^3 = x^3 - x + \dfrac{1}{3x} - \dfrac{1}{27x^3}$.

Therefore coefficient of x^3 is $\frac{1}{3} - \frac{4}{3} + \frac{2}{3} = -\frac{1}{3}$.

4.4 $(x + y)^5 = x^5 + 5x^4y + 10x^3y^2 + 10x^2y^3 + 5xy^4 + y^5.$

So $x^5 + y^5 = (x + y)^5 - 5xy(x^3 + y^3) - 10x^2y^2(x + y).$

But $x^3 + y^3 = (x + y)^3 - 3xy(x + y) = p^3 - 3qp$

$\Rightarrow x^5 + y^5 = p^5 - 5q(p^3 - 3pq) - 10q^2p = p^5 - 5p^3q + 5pq^2.$

4.5 $\left(1 + 3x + \dfrac{3x^2}{2}\right)^4 = 1 + 4\left(3x + \dfrac{3x^2}{2}\right) + 6\left(3x + \dfrac{3x^2}{2}\right)^2 + 4\left(3x + \dfrac{3x^2}{2}\right)^3 + \ldots,$

$(1 + ax)^n = 1 + nax + \dfrac{n(n-1)}{(1)(2)}a^2x^2 + \dfrac{n(n-1)(n-2)}{(1)(2)(3)}a^3x^3 + \ldots.$

Equating coefficients of x: $12 = na.$

Equating coefficients of x^2: $6 + (6)(9) = \dfrac{n(n-1)a^2}{2}.$

Solving: $a = 2, n = 6.$

x^3 coefficients are respectively 162 and 160. Difference = 2.

4.6

(a) $(1 + 2x + 3x^2)^5 = 1 + 5(2x + 3x^2) + 10(2x + 3x^2)^2 + 10(2x + 3x^2)^3 + \ldots$

$\qquad = 1 + 10x + 15x^2 + 40x^2 + 120x^3 + \ldots + 80x^3 + \ldots$

$\qquad = 1 + 10x + 55x^2 + 200x^3 + \ldots.$

(b) $\left(x^2 + \dfrac{1}{x}\right)^{12} = x^{24} + {}_{12}C_1x^{21} + {}_{12}C_2x^{18} + {}_{12}C_3x^{15} + \ldots.$

The term independent of x has coefficient ${}_{12}C_8 = \dfrac{12!}{8!\ 4!} = 495.$

4.7 $\dfrac{1}{1-x} + \dfrac{2}{1+2x} + \dfrac{6}{(1+2x)^2} = \dfrac{(1+2x)^2 + 2(1-x)(1+2x) + 6(1-x)}{(1-x)(1+2x)^2}$

$\qquad\qquad\qquad\qquad = \dfrac{9}{(1-x)(1+2x)^2}.$

Now, $\dfrac{1}{(1-x)} = 1 + x + x^2 + x^3 + \ldots$ (provided $|x| < 1$),

$\dfrac{2}{(1+2x)} = 2(1 - 2x + 4x^2 - 8x^3 + \ldots)$ (provided $|2x| < 1$),

and $\dfrac{6}{(1+2x)^2} = 6(1 - 4x + 12x^2 - 32x^3 + \ldots).$

Hence

$\qquad f(x) = 9 - 27x + 81x^2 + \text{higher powers of } x$ (if $|2x| < 1$, i.e. $|x| < \tfrac{1}{2}$).

4.8

$(1 + 3x^3)^{1/3} = 1 + \dfrac{1}{3}(3x^3) + \left(\dfrac{1}{3}\right)\left(-\dfrac{2}{3}\right)\dfrac{(3x^3)^2}{2} + \dfrac{1}{3}\left(-\dfrac{2}{3}\right)\left(-\dfrac{5}{3}\right)\dfrac{(3x^3)^3}{(2)(3)} + \ldots$

$\qquad = 1 + x^3 - x^6 + \dfrac{5}{3}x^9 - \ldots$ (valid if $|3x^3| < 1$).

Put $x = 0.2 = \dfrac{1}{5}$, $\left(1 + \dfrac{3}{125}\right)^{1/3} = \left(\dfrac{128}{125}\right)^{1/3} = \left(\dfrac{2^7}{5^3}\right)^{1/3} = \dfrac{4}{5}\sqrt[3]{2}$.

But from the series,

$$\left(1 + \dfrac{3}{125}\right)^{1/3} = 1 + (0.2)^3 - (0.2)^6 + \dfrac{5}{3}(0.2)^9 - \dots$$

$$= 1 + 0.008 - 0.000\,064 + 0.000\,000\,853 - \dots$$

$$= 1.007\,936\,9.$$

Thus $\sqrt[3]{2} = \dfrac{5}{4}(1.007\,936\,9) = 1.259\,92$.

Work to two more places of accuracy than required.

4.9

$$(1 + x)^{1/2} = 1 + \dfrac{1}{2}x + \left(\dfrac{1}{2}\right)\left(-\dfrac{1}{2}\right)\dfrac{x^2}{2} + \left(\dfrac{1}{2}\right)\left(-\dfrac{1}{2}\right)\left(-\dfrac{3}{2}\right)\dfrac{x^3}{6} + \dots,$$

$$(1 - x)^{-1/2} = 1 + \left(-\dfrac{1}{2}\right)(-x) + \left(-\dfrac{1}{2}\right)\left(-\dfrac{3}{2}\right)\dfrac{(-x)^2}{2} + \left(-\dfrac{1}{2}\right)\left(-\dfrac{3}{2}\right)\left(-\dfrac{5}{2}\right)\dfrac{(-x)^3}{6} + \dots.$$

So

$$(1 + x)^{1/2}(1 - x)^{-1/2} = \left(1 + \dfrac{x}{2} - \dfrac{x^2}{8} + \dfrac{x^3}{16} + \dots\right)\left(1 + \dfrac{x}{2} + \dfrac{3x^2}{8} + \dfrac{5x^3}{16} + \dots\right).$$

$$= 1 + x + \dfrac{x^2}{2} + \dfrac{x^3}{2} + \dots \qquad \text{(if } |x| < 1).$$

Try $x = \dfrac{1}{9}$. Then

$$\left(\dfrac{1 + x}{1 - x}\right)^{1/2} = \left(\dfrac{10}{8}\right)^{1/2} = \dfrac{1}{2}\sqrt{5}.$$

In the series $\dfrac{1}{2}\sqrt{5} = 1 + \dfrac{1}{9} + \dfrac{1}{162} + \dfrac{1}{1458} + \dots$

Hence $\qquad \sqrt{5} \approx \dfrac{1630}{729}$.

4.10

(a) $\qquad (1 + x)^{2n} = c_0 + c_1 x + \dots + c_{2n}x^{2n}$.

Let $x = 1$, $\quad 2^{2n} = c_0 + c_1 + \dots + c_{2n}$.

Let $x = -1$, $\quad 0 = c_0 - c_1 + \dots + c_{2n}$.

Add and divide by 2: $\quad 2^{2n-1} = c_0 + c_2 + \dots + c_{2n}$.

(b) $\qquad\qquad (1 + x)^n = c_0 + c_1 x + c_2 x^2 + \dots + c_n x^n$. $\hfill (1)$

Differentiate:

$$n(1 + x)^{n-1} = c_1 + 2c_2 x + 3c_3 x^2 + \dots + nc_n x^{n-1}. \hfill (2)$$

Let $x = 1$ in equations 1 and 2 and add:

$$2^n + (n)\,2^{n-1} = c_0 + 2c_1 + 3c_2 + \dots + (n + 1)c_n.$$

(c) Integrate:

$$\dfrac{(1 + x)^{n+1}}{n + 1} = c_0 x + c_1\dfrac{x^2}{2} + c_2\dfrac{x^3}{3} + \dots + \dfrac{c_n x^{n+1}}{n + 1} + K.$$

Let $x = 0 \Rightarrow K = \dfrac{1}{n + 1}$.

Let $x = -1 \Rightarrow 0 = -c_0 + \dfrac{c_1}{2} - \dfrac{c_2}{3} + \ldots + \dfrac{(-1)^{n+1}c_n}{n+1} + \dfrac{1}{n+1}$.

Hence $c_0 - \dfrac{c_1}{2} + \dfrac{c_2}{3} + \ldots + \dfrac{(-1)^n c_n}{n+1} = \dfrac{1}{n+1}$.

5 Partial Fractions

Partial fractions, to include denominators such as
$$(ax + b)(cx + d)(ex + f), (ax + b)(cx + d)^2 \text{ and } (ax + b)(x^2 + c^2).$$

5.1 Fact Sheet

If the degree of the numerator is equal to or higher than the degree of the denominator, the numerator must be divided by the denominator until the remainder is of lower degree than the denominator.

(a) Linear Factors

$$\frac{f(x)}{(ax + b)(cx + d)(ex + f)} \equiv \frac{A}{(ax + b)} + \frac{B}{(cx + d)} + \frac{C}{(ex + f)}.$$

(b) Repeated Factors

$$\frac{f(x)}{(ax + b)(cx + d)^2} \equiv \frac{A}{(ax + b)} + \frac{B}{(cx + d)} + \frac{C}{(cx + d)^2}.$$

(c) Quadratic Factors

$$\frac{f(x)}{(ax + b)(x^2 + c^2)} \equiv \frac{A}{(ax + b)} + \frac{Bx + C}{(x^2 + c^2)}.$$

To find the constants A, B and C. Put all the terms on the right-hand side over the same denominator as the left-hand side. Equate the numerators. Equate the coefficients of the powers of x and/or substitute well chosen values for x.

Partial fractions are frequently combined with the binomial series, differentiation (page 151) or integration (page 177).

Particular binomial series:

$(1 + x)^{-1} = 1 - x + x^2 - x^3 + x^4 - \ldots$ valid for $-1 < x < 1 \Leftrightarrow |x| < 1$,

$(1 + x)^{-2} = 1 - 2x + 3x^2 - 4x^3 + \ldots$ valid for $-1 < x < 1 \Leftrightarrow |x| < 1$.

5.2 Worked Examples

5.1 Express $\dfrac{3 + x}{(2 - x)(1 + 2x)}$ in partial fractions, and hence, or otherwise, obtain the first three non-zero terms in the expansion of this expression in ascending powers of x.

State the range of values of x for which the expansion is valid.

● **Look at the denominator—both factors are linear, need numerators A and B.**

$$f(x) = \frac{3 + x}{(2 - x)(1 + 2x)} \equiv \frac{A}{2 - x} + \frac{B}{1 + 2x}$$

$$\equiv \frac{A(1 + 2x) + B(2 - x)}{(2 - x)(1 + 2x)}.$$

Equating the numerators: $\qquad 3 + x \equiv A(1 + 2x) + B(2 - x).$

Put $x = 2$: $\qquad\qquad\qquad\qquad 5 = 5A \;\Rightarrow A = 1.$

Put $x = -\tfrac{1}{2}$: $\qquad\qquad\qquad 2\tfrac{1}{2} = 2\tfrac{1}{2}B \Rightarrow B = 1.$

Hence $f(x) \equiv \dfrac{3 + x}{(2 - x)(1 + 2x)} \equiv \dfrac{1}{2 - x} + \dfrac{1}{1 + 2x}$

$$\equiv \frac{1}{2(1 - x/2)} + \frac{1}{1 + 2x}$$

$$\equiv \tfrac{1}{2}(1 - x/2)^{-1} + (1 + 2x)^{-1}.$$

By the binomial expansion:

$$\tfrac{1}{2}(1 - x/2)^{-1} = \tfrac{1}{2}(1 + x/2 + (x/2)^2 + (x/2)^3 + \ldots)$$
$$\text{(valid for } -1 < x/2 < 1 \;\Rightarrow\; -2 < x < 2),$$
$$(1 + 2x)^{-1} = 1 - 2x + (2x)^2 - (2x)^3 + \ldots$$
$$\text{(valid for } -1 < 2x < 1 \;\Rightarrow\; -\tfrac{1}{2} < x < \tfrac{1}{2}).$$

Adding gives:
$$f(x) = \tfrac{1}{2} + x/4 + x^2/8 + x^3/16 + 1 - 2x + 4x^2 - 8x^3 + \ldots$$
$$= 3/2 - 7x/4 + 33x^2/8 \quad \text{to the first three terms.}$$

Validity is given when both conditions are valid—the smaller interval is taken.

Series is valid for $-\tfrac{1}{2} < x < \tfrac{1}{2}$.

5.2 Express $\dfrac{x^3 + 2x^2 - x + 3}{(x + 2)(x - 3)}$ in partial fractions.

● **Compare the orders of the numerator and denominator. Since the numerator is of order 3 and the denominator of order 2 a long division must be carried out first:**

$$
\begin{array}{r}
x + 3 \\
x^2 - x - 6 \overline{)\,x^3 + 2x^2 - x + 3} \\
\underline{x^3 - x^2 - 6x} \\
3x^2 + 5x + 3 \\
\underline{3x^2 - 3x - 18} \\
8x + 21.
\end{array}
$$

$$\frac{x^3 + 2x^2 - x + 3}{(x+2)(x-3)} = x + 3 + \frac{8x + 21}{(x+2)(x-3)} .$$

Let
$$\frac{8x + 21}{(x+2)(x-3)} \equiv \frac{A}{x+2} + \frac{B}{x-3}$$

$$\equiv \frac{A(x-3) + B(x+2)}{(x+2)(x-3)} .$$

Equating the numerators:
$$8x + 21 \equiv A(x-3) + B(x+2).$$

Put $x = 3$: $\qquad 45 = B(5), \qquad B = 9.$

Put $x = -2$: $\qquad 5 = A(-5), \qquad A = -1.$

Therefore
$$\frac{x^3 + 2x^2 - x + 3}{(x+2)(x-3)} \equiv x + 3 - \frac{1}{x+2} + \frac{9}{x-3} .$$

5.3 Express $x^3 - 3x^2 - 18x + 40$ as the product of three linear factors.

Hence express $\dfrac{9x - 72}{x^3 - 3x^2 - 18x + 40}$ in the form of

$\dfrac{A}{x-a} + \dfrac{B}{x-b} + \dfrac{C}{x-c}$, where a, b, c, A, B, C are numbers to be determined.

- **Find the first factor by the factor theorem.**

Let $f(x) = x^3 - 3x^2 - 18x + 40.$
Trying the factors of 40 as values of x:

$$f(1) = 1 - 3 - 18 + 40 \neq 0 \quad \Rightarrow \quad (x-1) \text{ is not a factor of } f(x).$$
$$f(2) = 8 - 12 - 36 + 40 = 0 \quad \Rightarrow \quad (x-2) \text{ is a factor of } f(x).$$
$$f(x) = (x-2)(x^2 - x - 20)$$
$$= (x-2)(x-5)(x+4).$$

Hence $x^3 - 3x^2 - 18x + 40 = (x-2)(x-5)(x+4).$
$$\frac{9x - 72}{(x-2)(x-5)(x+4)} \equiv \frac{A}{x-2} + \frac{B}{x-5} + \frac{C}{x+4}$$

$$\equiv \frac{A(x-5)(x+4) + B(x-2)(x+4) + C(x-2)(x-5)}{(x-2)(x-5)(x+4)} .$$

Equating the numerators:
$$9x - 72 \equiv A(x-5)(x+4) + B(x-2)(x+4) + C(x-2)(x-5).$$

Put $x = 2$: $\quad 18 - 72 = A(-3)(6), \quad -54 = -18A; \quad A = 3.$

Put $x = -4$: $\;-36 - 72 = C(-6)(-9), -108 = 54C; \quad C = -2.$

Put $x = 5$: $\quad 45 - 72 = B(3)(9), \qquad -27 = 27B; \quad B = -1.$

Hence
$$\frac{9x - 72}{x^3 - 3x^2 - 18x + 40} = \frac{3}{x-2} - \frac{1}{x-5} - \frac{2}{x+4} .$$

5.4 Express $f(x) = \dfrac{4x^2 - 7x + 3}{(2-x)(1+x^2)}$ in partial fractions.

Expand $f(x)$ in ascending powers of x as far as, and including, the term in x^3.
For what values of x is this expansion valid?

- Look at the denominator: one factor is linear and needs a numerator A, one is quadratic and needs a numerator $Bx + C$.

$$f(x) = \frac{4x^2 - 7x + 3}{(2 - x)(1 + x^2)} \equiv \frac{A}{2 - x} + \frac{Bx + C}{1 + x^2}$$

$$\equiv \frac{A(1 + x^2) + (Bx + C)(2 - x)}{(2 - x)(1 + x^2)}.$$

Equating the numerators:

$$4x^2 - 7x + 3 \equiv A(1 + x^2) + (Bx + C)(2 - x).$$

Put $x = 2$: $16 - 14 + 3 = A(1 + 4)$ $\Rightarrow$ $5A = 5, A = 1.$

Put $x = 0$: $3 = A(1) + C(2)$ $\Rightarrow$ $C = 1.$

Put $x = 1$: $4 - 7 + 3 = A(1 + 1) + (B + C)(1)$

$$0 = 2A + B + C \Rightarrow B = -3.$$

Alternatively, compare coefficients of x^2: $4 = A - B$ $\Rightarrow$ $B = -3.$

Hence $f(x) \equiv \dfrac{1}{(2 - x)} + \dfrac{1 - 3x}{(1 + x^2)}$

$$\frac{1}{2 - x} = \frac{1}{2(1 - x/2)} = \frac{1}{2}\left(1 - \frac{x}{2}\right)^{-1} = \frac{1}{2}\left(1 + \frac{x}{2} + \frac{x^2}{4} + \frac{x^3}{8} + \ldots\right)$$

(valid $-2 < x < 2$),

$$\frac{1}{1 + x^2} = (1 + x^2)^{-1} = 1 - x^2 + x^4 - \ldots$$

(valid $-1 < x < 1$).

$$f(x) = \frac{1}{2}\left(1 + \frac{x}{2} + \frac{x^2}{4} + \frac{x^3}{8} \ldots\right) + (1 - 3x)(1 - x^2 + x^4 - \ldots)$$

$$= \frac{1}{2} + \frac{x}{4} + \frac{x^2}{8} + \frac{x^3}{16} + 1 - 3x - x^2 + 3x^3 \ldots$$

$$= \frac{3}{2} - \frac{11}{4}x - \frac{7}{8}x^2 + \frac{49}{16}x^3 \ldots.$$

The series is valid when $-2 < x < 2$ and $-1 < x < 1$.
Taking the smaller range, $-1 < x < 1$.

5.5 Expand $\dfrac{16x^2 + 8x}{(1 + x)(1 + 3x)(1 + 5x)}$ in ascending powers of x, up to and including the term in x^3.

- All factors of the denominator are linear — need numerators A, B and C.

$$f(x) = \frac{16x^2 + 8x}{(1 + x)(1 + 3x)(1 + 5x)} \equiv \frac{A}{(1 + x)} + \frac{B}{(1 + 3x)} + \frac{C}{(1 + 5x)}$$

$$\equiv \frac{A(1 + 3x)(1 + 5x) + B(1 + x)(1 + 5x) + C(1 + x)(1 + 3x)}{(1 + x)(1 + 3x)(1 + 5x)}.$$

Equating the numerators:

$$16x^2 + 8x \equiv A(1 + 3x)(1 + 5x) + B(1 + x)(1 + 5x) + C(1 + x)(1 + 3x).$$

Put $x = -1$: $16 - 8 = A(-2)(-4)$ $\Rightarrow$ $A = 1.$

Put $x = -\frac{1}{3}$: $\frac{16}{9} - \frac{8}{3} = B(\frac{2}{3})(-\frac{2}{3}), -8 = -4B$ $\Rightarrow$ $B = 2.$

Put $x = -\frac{1}{5}$: $\frac{16}{25} - \frac{8}{5} = C(\frac{4}{5})(\frac{2}{5}), -24 = 8C$ $\Rightarrow$ $C = -3.$

$$f(x) = \frac{1}{1+x} + \frac{2}{(1+3x)} - \frac{3}{(1+5x)}$$

$$= (1+x)^{-1} + 2(1+3x)^{-1} - 3(1+5x)^{-1}$$

$$= (1 - x + x^2 - x^3 \ldots) + 2[1 - 3x + (3x)^2 - (3x)^3 \ldots]$$

$$\qquad - 3[1 - (5x) + (5x)^2 - (5x)^3 \ldots]$$

$$= 1 - x + x^2 - x^3 \ldots + 2 - 6x + 18x^2 - 54x^3 \ldots$$

$$\qquad \ldots - 3 + 15x - 75x^2 + 375x^3 \ldots$$

$$f(x) = 8x - 56x^2 + 320x^3 + \ldots$$

5.6 Express the function $\dfrac{9x}{(1+x)(1-2x)^2}$ as the sum of three partial fractions. Hence, or otherwise, find the first three terms in the expansion of the function in ascending powers of x.

- **Notice that one of the factors is a repeated factor $\Rightarrow$ there must be three fractions, denominators $(1+x)$, $(1-2x)$ and $(1-2x)^2$.**

Let
$$\frac{9x}{(1+x)(1-2x)^2} \equiv \frac{A}{(1+x)} + \frac{B}{(1-2x)} + \frac{C}{(1-2x)^2}$$

$$\equiv \frac{A(1-2x)^2 + B(1+x)(1-2x) + C(1+x)}{(1+x)(1-2x)^2}.$$

Equating the numerators:
$$9x \equiv A(1-2x)^2 + B(1+x)(1-2x) + C(1+x).$$

Put $x = -1$: $\quad -9 = A(3)^2 \qquad \Rightarrow A = -1.$
Put $x = \frac{1}{2}$: $\quad \frac{9}{2} = C(\frac{3}{2}) \qquad \Rightarrow C = 3.$
Put $x = 0$: $\quad 0 = A + B + C \qquad \Rightarrow B = -2.$

Hence
$$\frac{9x}{(1+x)(1-2x)^2} \equiv \frac{3}{(1-2x)^2} - \frac{1}{(1+x)} - \frac{2}{(1-2x)}.$$

$$\frac{3}{(1-2x)^2} = 3(1-2x)^{-2}$$

$$= 3[1 - 2(-2x) + 3(-2x)^2 - 4(-2x)^3 + \ldots]$$

$$= 3 + 12x + 36x^2 + 96x^3 + \ldots$$

$$\frac{1}{(1+x)} = (1+x)^{-1}$$

$$= 1 - x + x^2 - x^3 + \ldots$$

$$\frac{2}{(1-2x)} = 2(1-2x)^{-1}$$

$$= 2[1 + 2x + (2x)^2 + (2x)^3 + \ldots]$$

$$= 2 + 4x + 8x^2 + 16x^3 + \ldots$$

Hence
$$\frac{9x}{(1+x)(1-2x)^2} = (3 + 12x + 36x^2 + 96x^3)$$

$$- (1 - x + x^2 - x^3)$$

$$- (2 + 4x + 8x^2 + 16x^3) + \ldots$$

$$= 9x + 27x^2 + 81x^3 + \ldots$$

5.7 Write $3 + \dfrac{1}{x-2} - \dfrac{4}{x+3}$ as a single fraction. Hence express

$f(x) = \dfrac{3x^2 - 7}{(x^2 + x - 6)}$ in the form $A + \dfrac{B}{x-2} + \dfrac{C}{x+3}$

stating the values of the constants A, B and C.

Sketch the graph of $f(x)$, paying particular attention to the values of $f(x)$ near $x = 2$, $x = -3$ and when $|x|$ is large.

● As a single fraction:

$$3 + \frac{1}{x-2} - \frac{4}{x+3} = \frac{3(x-2)(x+3) + 1(x+3) - 4(x-2)}{(x-2)(x+3)}$$

$$= \frac{3x^2 + 3x - 18 + x + 3 - 4x + 8}{(x-2)(x+3)}$$

$$= \frac{3x^2 - 7}{(x-2)(x+3)}.$$

Hence
$$f(x) = \frac{3x^2 - 7}{(x^2 + x - 6)} = \frac{3x^2 - 7}{(x-2)(x+3)}$$

$$= 3 + \frac{1}{(x-2)} - \frac{4}{(x+3)}.$$

$A = 3$, $B = 1$, $C = -4$.

Stages in curve sketching:
(a) Look at denominator. When this is zero the function is undefined and there is an asymptote.
(b) Put numerator = 0. This gives points of intersection on the x-axis.
(c) Let $|x| \to \infty$.
(d) Find the point(s) of intersection with the y-axis.

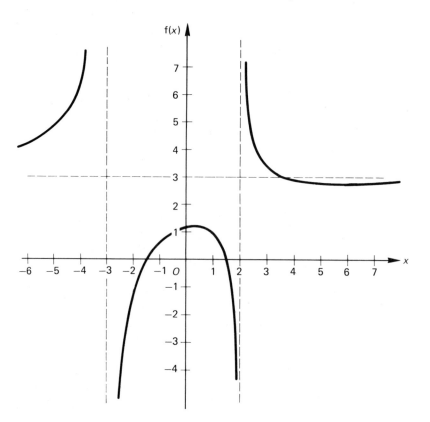

(a) $f(x)$ is not defined when $x = 2$ and -3. Two asymptotes.
When $x = 2.1$, $f(x) \approx 12.2$, which is large and positive.
When $x = -3.1$, $f(x) \approx 42.8$, which is large and positive.

(b) $3x^2 - 7 = 0$, $\qquad x = \pm\sqrt{(7/3)}$.

(c) As $|x| \to \infty$ $\quad f(x) \to 3$. Trying a large value of $|x|$ shows that $f(x) < 3$
when $x \to \infty$ $\quad$ and $f(x) > 3$ when $x \to -\infty$;

(d) When $x = 0$, $y = \frac{7}{6}$.

5.8 Given that $g(x) = \dfrac{5 - 5x}{(1 + x^2)(3 - x)}$, express $g(x)$ in partial fractions. Hence
or otherwise, show that the expansion of $g(x)$ as a series in ascending powers
of x, up to and including the term in x^4 is

$$\tfrac{5}{3} - \tfrac{10}{9}x - \tfrac{55}{27}x^2 + \tfrac{80}{81}x^3 + \tfrac{485}{243}x^4.$$

• **Notice that one of the factors of the denominator is a quadratic. Put the
numerator as $Ax + B$.**

Let $\qquad \dfrac{5 - 5x}{(1 + x^2)(3 - x)} \equiv \dfrac{Ax + B}{1 + x^2} + \dfrac{C}{3 - x}$

$$\equiv \dfrac{(Ax + B)(3 - x) + C(1 + x^2)}{(1 + x^2)(3 - x)}.$$

Equating the numerators: $5 - 5x \equiv (Ax + B)(3 - x) + C(1 + x^2)$.
Put $x = 3$: $\quad 5 - 15 = C(1 + 9)$, $\qquad \Rightarrow \quad C = -1$.
Put $x = 0$: $\quad 5 = B(3) + C$ $\qquad\qquad \Rightarrow \quad B = 2$.
Put $x = 1$: $\quad 0 = (A + B)(2) + C(2)$ $\Rightarrow \quad A = -1$.

Hence $\qquad\qquad \dfrac{5 - 5x}{(1 + x^2)(3 - x)} = \dfrac{2 - x}{1 + x^2} - \dfrac{1}{3 - x}$.

$$\dfrac{2 - x}{1 + x^2} = (2 - x)(1 + x^2)^{-1} = (2 - x)(1 - x^2 + x^4 \ldots)$$

$$= 2 - x - 2x^2 + x^3 + 2x^4 \ldots.$$

$$\dfrac{1}{3 - x} = \dfrac{1}{3(1 - x/3)} = \tfrac{1}{3}(1 - x/3)^{-1}$$

$$= \tfrac{1}{3}(1 + x/3 + x^2/9 + x^3/27 + x^4/81 + \ldots)$$

$$= \tfrac{1}{3} + x/9 + x^2/27 + x^3/81 + x^4/243 + \ldots.$$

Hence
$g(x) = 2 - x - 2x^2 + x^3 + 2x^4 - \tfrac{1}{3} - x/9 - x^2/27 - x^3/81 - x^4/243 - \ldots$
$= \tfrac{5}{3} - \tfrac{10}{9}x - \tfrac{55}{27}x^2 + \tfrac{80}{81}x^3 + \tfrac{485}{243}x^4 + \ldots.$

5.3 Exercises

5.1 Express $\dfrac{1}{(1 - x)(1 + 2x)}$ in partial fractions. Hence find the first three terms

of the expansion of $\dfrac{1}{(1 - x)(1 + 2x)}$ in ascending powers of x.

Find the coefficient of x^n and state the range of values of x for which the expansion is valid.

5.2 Given that $f(x) \equiv \dfrac{7}{(3x - 1)(x + 2)}$, express $f(x)$ in partial fractions.

Sketch the curve $y = f(x)$, showing the asymptotes and the points of intersection of the curve with the axes.

5.3 Express $\dfrac{36 - 2x}{(2x + 1)(9 + x^2)}$ in partial fractions.

5.4 Given that $f(x) \equiv \dfrac{2x^2 + x - 43}{(x + 3)(x - 4)}$, express $f(x)$ in partial fractions and hence find the first three terms when $f(x)$ is expanded in a series of ascending powers of x, stating the set of values for which the expansion is valid.

5.5 Given that $g(x) \equiv \dfrac{3}{(1 + x)^2 (1 + 2x)}$, express $g(x)$ in partial fractions and hence find the first three terms in the series when $g(x)$ is expanded in a series of ascending powers of $\dfrac{1}{x}$, stating the set of values for which the expansion is valid.

5.6 Express $\dfrac{1}{r(r + 1)}$ in partial fractions.

Hence or otherwise find $S_n = \displaystyle\sum_{r=1}^{n} \dfrac{1}{r(r + 1)}$ and deduce the value of S_n as n tends to infinity.

5.7 Express $\dfrac{1}{r^2 - 1}$ in partial fractions.

Determine $\displaystyle\sum_{r=2}^{n} \dfrac{1}{r^2 - 1}$ and hence deduce the sum to infinity.

5.4 Outline Solutions to Exercises

5.1 $\dfrac{1}{(1 - x)(1 + 2x)} = \dfrac{1}{3(1 - x)} + \dfrac{2}{3(1 + 2x)}$,

$\dfrac{1}{3(1 - x)} = \tfrac{1}{3}(1 - x)^{-1} = \tfrac{1}{3}(1 + x + x^2 + \ldots + x^n \ldots)$ (if $|x| < 1$),

$\dfrac{2}{3(1 + 2x)} = \tfrac{2}{3}(1 + 2x)^{-1} = \tfrac{2}{3}[1 - 2x + 4x^2 - \ldots + (-1)^n (2x)^n + \ldots]$ (if $|x| < \tfrac{1}{2}$).

Hence
$$f(x) = \tfrac{1}{3}[1 + x + x^2 + \ldots + x^n + \ldots + 2 - 4x + 8x^2 - \ldots + 2(-2x)^n \ldots]$$
$$= \tfrac{1}{3}\{3 - 3x + 9x^2 + \ldots + [1 - (-2)^{n+1}]x^n + \ldots\}.$$
$$= 1 - x + 3x^2 + \ldots .$$

Coefficient of $x^n = \tfrac{1}{3}[1 - (-2)^{n+1}]$ (valid if $-\tfrac{1}{2} < x < \tfrac{1}{2}$).

44

5.2 $\dfrac{7}{(3x-1)(x+2)} = \dfrac{3}{(3x-1)} - \dfrac{1}{(x+2)}$.

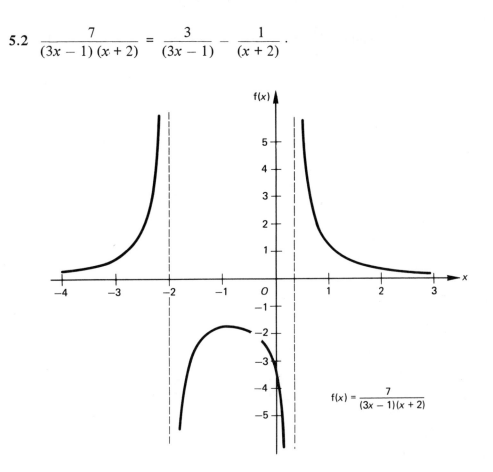

$$f(x) = \dfrac{7}{(3x-1)(x+2)}$$

5.3 $\dfrac{36-2x}{(2x+1)(9+x^2)} = \dfrac{A}{2x+1} + \dfrac{Bx+C}{9+x^2} = \dfrac{4}{2x+1} - \dfrac{2x}{9+x^2}$.

5.4 $f(x) = \dfrac{2x^2+x-43}{(x+3)(x-4)} = 2 + \dfrac{3x-19}{(x+3)(x-4)}$

$$= 2 + \dfrac{4}{x+3} - \dfrac{1}{x-4}$$

Now

$$\dfrac{4}{x+3} = \dfrac{4}{3}\left(1+\dfrac{x}{3}\right)^{-1} = \dfrac{4}{3}\left(1 - \dfrac{x}{3} + \dfrac{x^2}{9} - \dfrac{x^3}{27} + \ldots\right) \qquad \left(\text{if } \left|\dfrac{x}{3}\right| < 1\right);$$

$$\dfrac{-1}{x-4} = \dfrac{1}{4}\left(1-\dfrac{x}{4}\right)^{-1} = \dfrac{1}{4}\left(1 + \dfrac{x}{4} + \dfrac{x^2}{16} + \dfrac{x^3}{64} + \ldots\right) \qquad \left(\text{if } \left|\dfrac{x}{4}\right| < 1\right).$$

Hence $f(x) = 2 + \dfrac{4}{3} + \dfrac{1}{4} + \left(-\dfrac{4}{9} + \dfrac{1}{16}\right)x + \left(\dfrac{4}{27} + \dfrac{1}{64}\right)x^2 + \ldots$

$$= \dfrac{43}{12} - \dfrac{55}{144}x + \dfrac{283}{1728}x^2 + \ldots \qquad (\text{for } -3 < x < 3).$$

5.5 $g(x) = \dfrac{-6}{1+x} - \dfrac{3}{(1+x)^2} + \dfrac{12}{1+2x}$.

To obtain series in ascending powers of $\left(\dfrac{1}{x}\right)$, express $\dfrac{1}{1+x}$ in the form $\dfrac{1}{x}\left(1 + \dfrac{1}{x}\right)^{-1}$ and then use the binomial.

Valid for $\left|\dfrac{1}{x}\right| < 1 \quad \Rightarrow \quad |x| > 1 \quad \Rightarrow \quad x < -1 \text{ or } x > 1.$

$$\xleftarrow{\qquad\text{Valid}\ \ \big|\ \ \text{Not valid}\ \ \big|\ \ \text{Valid}\qquad}\rightarrow$$
$$\underset{-1}{}\qquad\underset{1}{}$$

$$\frac{-6}{1+x} = \frac{-6}{x}\left(1+\frac{1}{x}\right)^{-1} = \frac{-6}{x}\left(1-\frac{1}{x}+\frac{1}{x^2}-\frac{1}{x^3}+\frac{1}{x^4}+\ldots\right)\quad\left(\text{for}\left|\frac{1}{x}\right|<1\right).$$

$$\frac{-3}{(1+x)^2} = \frac{-3}{x^2}\left(1+\frac{1}{x}\right)^{-2} = \frac{-3}{x^2}\left(1-\frac{2}{x}+\frac{3}{x^2}-\frac{4}{x^3}+\ldots\right)\quad\left(\text{for}\left|\frac{1}{x}\right|<1\right),$$

$$\frac{12}{1+2x} = \frac{12}{2x}\left(1+\frac{1}{2x}\right)^{-1} = \frac{6}{x}\left(1-\frac{1}{2x}+\frac{1}{4x^2}-\frac{1}{8x^3}+\frac{1}{16x^4}\cdots\right)\left(\text{for}\left|\frac{1}{2x}\right|<1\right)$$

Hence $g(x) = \dfrac{3}{2x^3} - \dfrac{15}{4x^4} + \dfrac{51}{8x^5} + \ldots \qquad \left(\text{valid for}\left|\dfrac{1}{x}\right|<1\right).$

5.6 $\quad \dfrac{1}{r(r+1)} = \dfrac{1}{r} - \dfrac{1}{r+1}.$

$$S_n = \sum_{r=1}^{n}\frac{1}{r} - \frac{1}{r+1} = \sum_{r=1}^{n}\frac{1}{r} - \sum_{r=2}^{n+1}\frac{1}{r} = \frac{1}{1} - \frac{1}{n+1}.$$

$$\text{As } n \to \infty,\ \frac{1}{n+1} \to 0, \text{ so } S_\infty = 1.$$

5.7 $\quad \dfrac{1}{r^2 - 1} = \dfrac{1}{(r-1)(r+1)} = \dfrac{1}{2}\left(\dfrac{1}{r-1} - \dfrac{1}{r+1}\right).$

$$S_n = \sum_{r=2}^{n}\frac{1}{r^2-1} = \frac{1}{2}\sum_{r=2}^{n}\left(\frac{1}{r-1}-\frac{1}{r+1}\right) = \frac{1}{2}\sum_{r=1}^{n-1}\frac{1}{r} - \frac{1}{2}\sum_{r=3}^{n+1}\frac{1}{r}$$

$$= \frac{1}{2}\left[\left(\frac{1}{1}+\frac{1}{2}\right)-\left(\frac{1}{n}+\frac{1}{n+1}\right)\right]$$

$$= \frac{1}{2}\left(\frac{3}{2}-\frac{1}{n}-\frac{1}{n+1}\right).$$

As $n \to \infty,\ S_\infty \to \frac{3}{4}.$

6 Inequalities

The manipulation of simple algebraic inequalities. The function $|x|$. The solution of inequalities reducible to the form $f(x) > 0$, where $f(x)$ can then be expressed in factors; sketches of the graphs of $y = f(x)$ in these cases.

6.1 Fact Sheet

(a) Notation: Inequalities

- If a and b are numbers, represented by points on the x-axis, and $a > b$, then a is to the right of b.
- Thus $6 > 3$ and $-6 > -9$.
- If $a \geqslant b$ then a is greater than b or equal to b.

$$b \quad < \quad a$$

(b) Modulus

- $|x|$ may be regarded as the distance from the origin to the point on the x-axis.
- $|2| = 2$, $|-3| = 3$, $|x| = 5 \Rightarrow x = 5$ or -5.
  $|x| < 5 \Rightarrow -5 < x < 5$.
- $|f(x)| = f(x)$ if $f(x)$ is positive and $-f(x)$ if $f(x)$ is negative.
- If $f(x) = x^2 - 4$ then $f(3) = 5$, $|f(3)| = 5$,
  $$f(1) = -3, \quad |f(1)| = 3.$$

(c) Properties of Inequalities

- Any quantity may be added to or subtracted from both sides of an inequality:
  If $a > b$ then $a + c > b + c$ and $a - c > b - c$.
- Both sides of an inequality may be multiplied or divided by any positive quantity:
  If $c > 0$ and $a > b$ then $ac > bc$ and $\dfrac{a}{c} > \dfrac{b}{c}$.
- If both sides of an inequality are multiplied or divided by a negative quantity then the inequality is reversed:
  If $d < 0$ and $a > b$ then $ad < bd$ and $\dfrac{a}{d} < \dfrac{b}{d}$.

  (Play safe — never multiply or divide by a negative number.)
- If inequalities are of the same kind they may be added together:
  If $a > b$ and $c > d$ then $a + c > b + d$.
- **Never subtract inequalities.**
  (Example: $5 > 2$ and $4 > 0$ but $5 - 4 \not> 2 - 0$.)

6.2 Worked Examples

6.1
(a) Sketch the graphs of (i) $y = 2|x - 3|$, (ii) $y = |x^2 - 4|$.
(b) Find, in each case, the set of real values of x for which
   (i) $2(x - 1) \geqslant x + 2$,   (ii) $2|x - 1| \geqslant |x + 2|$.

● (a)

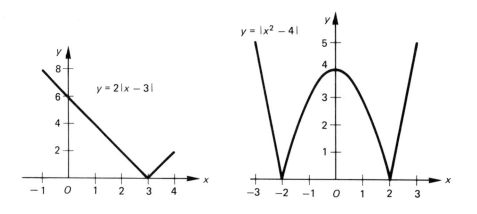

(b) (i) **Sketch $y_1 = 2(x - 1)$ and $y_2 = x + 2$ and find the range of values for which $y_1 \geqslant y_2$.**
   From the sketch it can be seen that there is just one range.

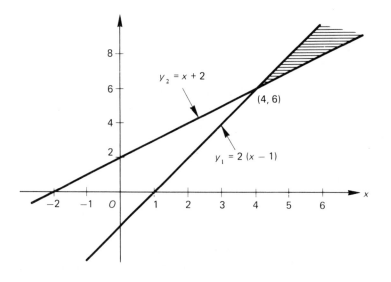

Finding the point of intersection of the lines:
$$y = 2x - 2 \qquad \text{and} \qquad y = x + 2,$$
$$\Rightarrow \qquad 2x - 2 = x + 2$$
$$\Rightarrow \qquad x = 4$$

$$2(x - 1) \geqslant x + 2 \qquad \text{for all } x \geqslant 4.$$

(ii) From the sketch of $y_1 = 2|x - 1|$ and $y_2 = |x + 2|$ it can be seen that the graphs intersect twice.
   Point A is given by $y_1 = 2(x - 1)$ and $y_2 = x + 2$.
   Solving gives $x = 4$ (from part (i)) $\Rightarrow y_1 \geqslant y_2$ when $x \geqslant 4$.
   Point B is given by $y_1 = -2(x - 1)$ and $y_2 = x + 2$.

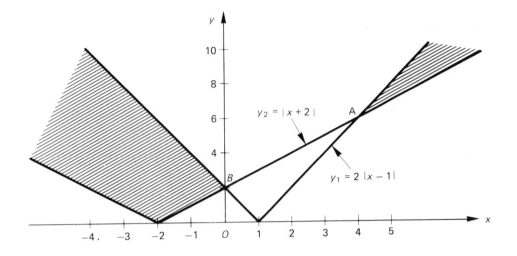

Finding the point of intersection of these lines:
$$-2x + 2 = x + 2$$
$$\Rightarrow \quad x = 0,$$
$$y_1 \geqslant y_2 \quad \text{when } x \leqslant 0.$$
Hence $2\,|x - 1| \geqslant |x + 2|$ when $x \leqslant 0$ or $x \geqslant 4$.

Alternatively, square both sides of the given inequality:
$$4(x - 1)^2 \geqslant (x + 2)^2$$
$$4x^2 - 8x + 4 \geqslant x^2 + 4x + 4$$
$$3x^2 - 12x \geqslant 0$$
$$3x\,(x - 4) \geqslant 0.$$
This gives $x \leqslant 0 \quad$ or $\quad x \geqslant 4$.

It is only permissible to square both sides of the inequality because each side is positive, or zero, for all x.

6.2 In one diagram sketch the three lines
$$x + y - 4 = 0, \qquad 2y - 3x - 3 = 0, \qquad 3x - y + 6 = 0.$$
Indicate, by shading in your diagram, the region in which the following three inequalities are all satisfied, marking this region A:
$$x + y - 4 < 0, \qquad 2y - 3x - 3 > 0, \qquad 3x - y + 6 < 0.$$

● **To find the required region substitute the coordinates of any point not on a line into the corresponding inequality. If the inequality is satisfied all points on the same side of the line as the chosen point will also satisfy the inequality. Lightly shade the other side of the line, i.e. the area which is not wanted. The origin is a useful point to try. See the figure on p. 50.**

$x + y - 4 < 0$ is satisfied by $(0, 0)$ so shade the region which does not contain $(0, 0)$.

$2y - 3x - 3 > 0$ is not satisfied by $(0, 0)$ so shade the region containing $(0, 0)$.

$3x - y + 6 < 0$ is not satisfied by $(0, 0)$ so shade the region containing $(0, 0)$.

The region which satisfies all the inequalities is unshaded, and is marked A.

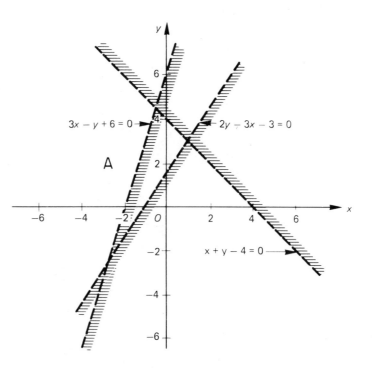

3x − y + 6 = 0→

←2y − 3x − 3 = 0

A

x + y − 4 = 0 →

6.3 Prove that for all real x, $0 < \dfrac{1}{x^2 + 10x + 27} \leqslant \frac{1}{2}$.

Sketch the curve $y = \dfrac{1}{x^2 + 10x + 27}$.

- The implications of $0 < \dfrac{1}{x^2 + 10x + 27} \leqslant \frac{1}{2}$ are

  (i) $x^2 + 10x + 27$ is always positive, and
  (ii) $x^2 + 10x + 27 \geqslant 2$ for all real x.

  These conditions are both satisfied if $x^2 + 10x + 27 \geqslant 2$ for all real x.

  Let $f(x) = x^2 + 10x + 27$
  $\qquad\quad = (x + 5)^2 + 2$.

  Since $(x + 5)^2 \geqslant 0$ for all real x, $f(x) \geqslant 2$ for all real x.

  Thus $0 < \dfrac{1}{x^2 + 10x + 27} \leqslant \frac{1}{2}$ for all real x.

  When $x = -5$, $f(x) = 2$, $\Rightarrow y = \frac{1}{2}$ (maximum point).

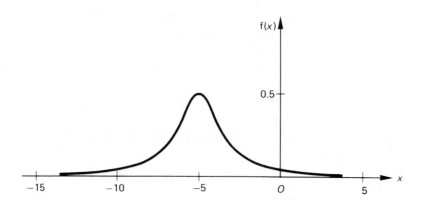

6.4

(a) Show that the arithmetic mean of two positive real numbers is greater than, or equal to, their geometric mean. Hence show that when w, x, y, z are real

$$w^4 + x^4 + y^4 + z^4 \geqslant 4wxyz.$$

(b) Find the solution set of the inequality

$$\frac{8}{x} < x + 2 \qquad (x \in \mathbb{R}, x \neq 0).$$

● (a) The arithmetic mean of w and x is $\dfrac{w + x}{2}$; the geometric mean is $\sqrt{(wx)}$, where $\sqrt{}$ means the positive square root.

In inequality questions it is often worth considering a bracket squared, for example $(w - x)^2$.

$(w - x)^2 \geqslant 0,$ for all real $w, x,$
$\Rightarrow \quad w^2 + x^2 - 2wx \geqslant 0.$
Add $4wx$ to each side:
$$w^2 + x^2 + 2wx \geqslant 4wx.$$
$$\Rightarrow \quad (w + x)^2 \geqslant 4wx.$$
Since w and x are both positive, take square roots

$$w + x \geqslant 2\sqrt{(wx)} \qquad \text{or} \qquad \frac{w + x}{2} \geqslant \sqrt{(wx)}$$

for all real positive values of x.
From this, $\quad w^4 + x^4 \geqslant 2\sqrt{(w^4 x^4)} = 2w^2 x^2 = 2(wx)^2$
$$y^4 + z^4 \geqslant 2\sqrt{(y^4 z^4)} = 2y^2 z^2 = 2(yz)^2.$$
Hence $w^4 + x^4 + y^4 + z^4 \geqslant 2[(wx)^2 + (yz)^2]$.
But $(wx)^2 + (yz)^2 \geqslant 2\sqrt{(w^2 x^2 y^2 z^2)} = 2wxyz$.
Hence $w^4 + x^4 + y^4 + z^4 \geqslant 4wxyz$.

(b) $\dfrac{8}{x} < x + 2.$

Do *not* multiply by x, since x may be negative.

Multiply by x^2 $8x < x^2 (x + 2)$
$$0 < x^3 + 2x^2 - 8x$$
$$0 < x(x + 4)(x - 2).$$
From the sketch it can be seen that
$$x > 2 \qquad \text{or} \qquad -4 < x < 0.$$

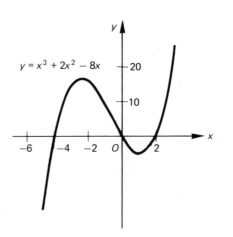

6.5 The set S is $\{(x, y) : 5x + y \geqslant 13 \text{ and } x^2 + y^2 \leqslant 13, (x, y) \in \mathbb{R} \times \mathbb{R}\}$.

Show clearly on a sketch the region in which the points representing the members of S must lie. If $(x, kx) \in S$ for at least one value of x, find the set of possible values of k.

- $5x + y = 13$ is a straight line.
  $x^2 + y^2 = 13$ is a circle centre $(0, 0)$.

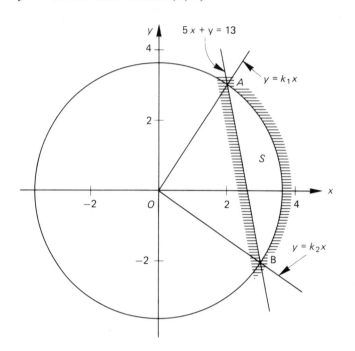

The points representing the members of S lie in the unshaded segment and include the perimeter. If A and B are the points of intersection of the line and the circle then

$$5x + y = 13 \tag{1}$$

and
$$x^2 + y^2 = 13 \tag{2}$$

at A and B.
From equation 1, $y = 13 - 5x$.
In equation 2,
$$x^2 + (13 - 5x)^2 = 13$$
$$26x^2 - 130x + 156 = 0$$
$$26 (x - 2)(x - 3) = 0$$

i.e. $x = 2$ or 3.
From equation 1,
$$x = 2 \quad \Rightarrow \quad y = 3 \qquad A \ (2, 3)$$
$$x = 3 \quad \Rightarrow \quad y = -2 \qquad B \ (3, -2).$$

All points (x, kx) lie on the line $y = kx$, so k is the gradient of the line joining the origin to any point of S.

If at least one point on the line $y = kx$ lies in S, then $y = kx$ must be between OA and OB.
Line OA has gradient $\frac{3}{2}$, OB has gradient $-\frac{2}{3}$.
Set of possible values of k is $-\frac{2}{3} \leqslant k \leqslant \frac{3}{2}$.

6.6
(a) For what values of x is $x^3 - 2x^2 > 5x - 6$?

(b) Find the set of real values of x for which $\dfrac{x(x - 3)}{(x - 2)} > 2$.

52

- (a) If $x^3 - 2x^2 > 5x - 6$ then $x^3 - 2x^2 - 5x + 6 > 0$.

  Let $f(x) = x^3 - 2x^2 - 5x + 6$.

  Use the factor theorem to find the factors of $f(x)$.

  Values of x to try are the factors of 6, (± 1, ± 2, ± 3, ± 6).

  $f(1) = 1 - 2 - 5 + 6 = 0 \quad$ so $(x - 1)$ is a factor of $f(x)$

  $f(2) = 8 - 8 - 10 + 6 = -4 \quad$ so $(x - 2)$ is not a factor.

  $f(3) = 27 - 18 - 15 + 6 = 0 \quad$ so $(x - 3)$ is a factor.

  $f(-2) = -8 - 8 + 10 + 6 = 0 \quad$ so $(x + 2)$ is a factor.

  $f(x) = (x - 1)(x - 3)(x + 2)$.

  Check that the coefficient of x^3 is correct, since the linear factors may be multiplied by a constant.

  If $f(x) > 0$, $(x - 1)(x - 3)(x + 2) > 0$.

  From the sketch or by considering the number line,
  $$x^3 - 2x^2 > 5x - 6 \qquad \text{when } -2 < x < 1 \text{ or } x > 3.$$

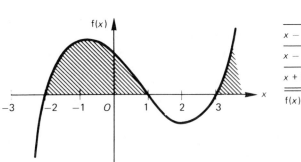

	-2	-1	0	1	2	3
$x - 1$	$-$		$-$		$+$	$+$
$x - 3$	$-$		$-$		$-$	$+$
$x + 2$	$-$		$+$		$+$	$+$
$f(x)$	$-$		$+$			$+$

  (b) If $\dfrac{x(x - 3)}{x - 2} > 2$ then $\dfrac{x(x - 3)}{x - 2} - 2 > 0$

  $$\frac{x(x - 3) - 2(x - 2)}{x - 2} > 0$$

  $$\frac{x^2 - 3x - 2x + 4}{x - 2} > 0$$

  $$\frac{(x - 4)(x - 1)}{x - 2} > 0.$$

  For a fraction to be positive either all the factors must be positive or an even number of factors negative for given values of x.

	0	1	2	3	4
$x - 4$	$-$		$-$	$-$	$+$
$x - 1$	$-$		$+$	$+$	$+$
$x - 2$	$-$		$-$	$+$	$+$
$f(x)$	$-$		$+$	$-$	$+$

  True if $x > 4$ or if $1 < x < 2$.

  The inequality must not be multiplied by $(x - 2)$ since for some values of x, $x - 2$ is negative.

  An alternative method is to multiply by $(x - 2)^2$, which is never negative for real values of x.

$$x(x-3)(x-2) > 2(x-2)^2$$
$$\Rightarrow \quad x(x-3)(x-2) - 2(x-2)^2 > 0$$
$$(x-2)(x^2 - 3x - 2x + 4) > 0$$
$$(x-2)(x^2 - 5x + 4) > 0$$
$$(x-2)(x-1)(x-4) > 0.$$

Use the same number line as for the first method.

6.7

(a) If $y = x^4 - 73x^2 + 888$ and $|y| \leqslant 312$, find the set of possible real values of x. Sketch the function $f(x) = x^4 - 73x^2 + 888$.

(b) Solve the inequality $\dfrac{1}{x-4} > \dfrac{1}{3-x}$.

● (a) $y = x^4 - 73x^2 + 888$ and $-312 \leqslant y \leqslant 312$.

(i) Substitute $y = 312$ into the equation
$$x^4 - 73x^2 + 888 = 312$$
$$\Rightarrow \quad x^4 - 73x^2 + 576 = 0.$$

This is a quadratic equation in x^2:
$$(x^2 - 9)(x^2 - 64) = 0 \quad \Rightarrow \quad x = \pm 3, x = \pm 8.$$

(ii) Substitute $y = -312$ into the equation
$$x^4 - 73x^2 + 888 = -312$$
$$\Rightarrow \quad x^4 - 73x^2 + 1200 = 0$$
$$\Rightarrow \quad x = \pm 5, x = \pm\sqrt{(48)} = \pm 4\sqrt{3}.$$

The function $f(x) = x^4 - 73x^2 + 888$ is an even quartic function so must be symmetrical about the line $x = 0$.

If $|y| \leqslant 312$ then $-8 \leqslant x \leqslant -4\sqrt{3}$ or $-5 \leqslant x \leqslant -3$ or $3 \leqslant x \leqslant 5$ or $4\sqrt{3} \leqslant x \leqslant 8$.

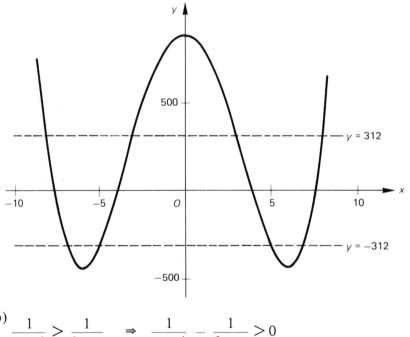

(b)

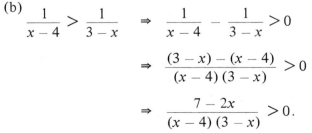

$$\dfrac{1}{x-4} > \dfrac{1}{3-x} \quad \Rightarrow \quad \dfrac{1}{x-4} - \dfrac{1}{3-x} > 0$$

$$\Rightarrow \quad \dfrac{(3-x)-(x-4)}{(x-4)(3-x)} > 0$$

$$\Rightarrow \quad \dfrac{7-2x}{(x-4)(3-x)} > 0.$$

		3.0	3.5	4.0	
$7 - 2x$	$+$		$+$	$-$	$-$
$x - 4$	$-$		$-$	$-$	$+$
$3 - x$	$+$		$-$	$-$	$-$
$f(x)$	$-$		$+$	$-$	$+$

Considering the number line,

$$\frac{1}{x-4} > \frac{1}{3-x} \quad \text{when } 3 < x < 3.5 \quad \text{or} \quad 4 < x.$$

6.3 Exercises

6.1 Find the sets of values of x for which
(a) $|3x - 5| < 6$;
(b) $(3x - 1)(x + 3) > 0$;
(c) $|x^2 - 9| < 8$.

6.2 Find the sets of values of x for which
(a) $|x - 3| > 2|x + 1|$,

(b) $\dfrac{x}{x + 2} < 2.$

6.3 In one diagram sketch the graphs $2x + 3y = 8$, $2x - 5y = -8$. Find the range of possible values for x, given that x and y satisfy the inequalities $2x + 3y \leqslant 8$, $5y - 2x > 8$.

6.4 Prove that for all real x, $\quad 0 < \dfrac{1}{x^2 - 5x + 9} \leqslant \dfrac{4}{11}.$

Sketch the curve $y = \dfrac{1}{x^2 - 5x + 9}.$

6.5 For what values of x is $f(x) = x^3 - 12x^2 + 39x - 28 < 0$?
Sketch the graph of $f(x)$

6.6 Find the set of values of x for which $f(x) > \tfrac{1}{2}$ where $f(x) = \dfrac{x(x - 2)}{(x + 3)}.$
Sketch the graph of $f(x)$.

6.7

(a) Show that, for all real values of x and y, $x^2 + y^2 \geqslant 2xy$.

(b) Hence show that, if a, b, c and d are real,

$$\left(\frac{a}{b}\right)^4 + \left(\frac{b}{c}\right)^4 + \left(\frac{c}{d}\right)^4 + \left(\frac{d}{a}\right)^4 \geqslant 4.$$

6.8 Prove that, for all real x, $0 \leqslant \dfrac{x^2 - 2x + 1}{x^2 + 4x + 5} \leqslant 10$.

Sketch the curve $y = \dfrac{x^2 - 2x + 1}{x^2 + 4x + 5}$.

6.4 Outline Solutions to Exercises

6.1

(a) $|3x - 5| < 6$

$\Rightarrow \quad 3x - 5 < 6 \qquad$ or $\qquad 5 - 3x < 6$

$\Rightarrow \quad x < \frac{11}{3} \qquad$ or $\qquad x > -\frac{1}{3}$.

Thus $-\frac{1}{3} < x < \frac{11}{3}$.

(b) $(3x - 1)(x + 3) > 0$.

From the sketch, $f(x) = (3x - 1)(x + 3)$, $f(x) > 0$ when $x < -3$ or $x > \frac{1}{3}$.

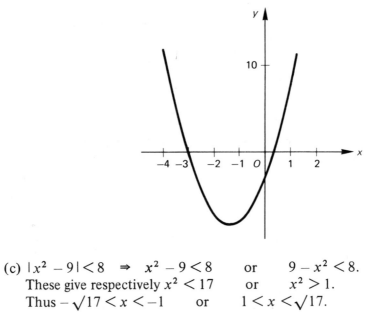

(c) $|x^2 - 9| < 8 \quad \Rightarrow \quad x^2 - 9 < 8 \qquad$ or $\qquad 9 - x^2 < 8$.

These give respectively $x^2 < 17 \qquad$ or $\qquad x^2 > 1$.

Thus $-\sqrt{17} < x < -1 \qquad$ or $\qquad 1 < x < \sqrt{17}$.

6.2

(a) $|x - 3| > 2\,|x + 1|$.

Squaring, $x^2 - 6x + 9 > 4x^2 + 8x + 4$

$\Rightarrow \quad 0 > (3x - 1)(x + 5)$.

	-5	$\frac{1}{3}$	
$3x - 1$	$-$	$-$	$+$
$x + 5$	$-$	$+$	$+$

Using the number line:

$|x - 3| > 2|x + 1|$ when $-5 < x < \frac{1}{3}$.

(b) $\dfrac{x}{x+2} < 2.$

  Multiply both sides by $(x+2)^2$:
  $$x(x+2) < 2(x+2)^2 \quad \Rightarrow \quad 0 < (x+2)(x+4).$$

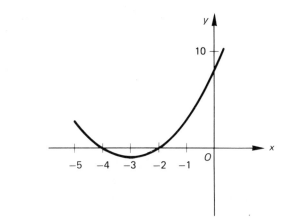

  So $x > -2$ or $x < -4.$

6.3 $2x + 3y \leqslant 8.$ Multiply by 5: $10x + 15y \leqslant 40.$
      $2x - 5y < -8.$ Multiply by 3: $6x - 15y < -24.$
                         Add: $16x \qquad < 16$

   $\Rightarrow x < 1.$

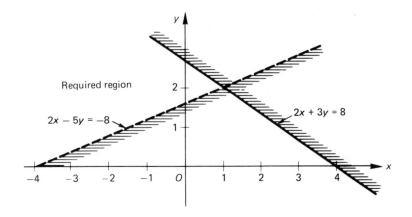

6.4 Let $y = \dfrac{1}{x^2 - 5x + 9} \quad \Rightarrow \quad yx^2 - 5xy + 9y - 1 = 0.$ (1)

  x is real, so $25y^2 \geqslant 4y(9y - 1) \quad \Rightarrow \quad 0 \geqslant y(11y - 4).$

  Thus $0 \leqslant y \leqslant \frac{4}{11}.$

  If $y = 0$, equation 1 gives $-1 = 0$, which is unacceptable, hence $y > 0.$
  If $y = \frac{4}{11}$, equation 1 gives $(2x - 5)^2 = 0$
  $\Rightarrow \quad x = \frac{5}{2}.$

  Hence $0 < \dfrac{1}{x^2 - 5x + 9} \leqslant \frac{4}{11}.$

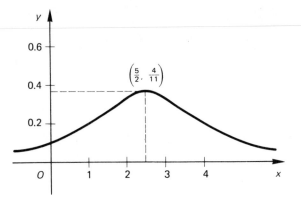

6.5 $f(x) = x^3 - 12x^2 + 39x - 28.$

Factor theorem gives $(x - 1)$ as a factor $\Rightarrow$ $f(x) = (x - 1)(x - 7)(x - 4).$

From the sketch, $f(x) < 0$ when $x < 1$ or $4 < x < 7.$

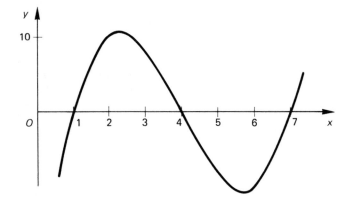

6.6 $\dfrac{x(x - 2)}{(x + 3)} > \tfrac{1}{2}.$ Multiply by $2(x + 3)^2$:

$$2x(x - 2)(x + 3) > (x + 3)^2.$$
$$\Rightarrow \quad (x + 3)(2x + 1)(x - 3) > 0.$$

Using the number line:

	-3		$-\tfrac{1}{2}$		3	
$x + 3$	$-$		$+$	$+$		$+$
$2x + 1$	$-$		$-$	$+$		$+$
$x - 3$	$-$		$-$	$-$		$+$
$(x + 3)(2x + 1)(x - 3)$	$-$		$+$	$-$		$+$

$$\dfrac{x(x - 2)}{(x + 3)} > \tfrac{1}{2} \qquad \text{when } -3 < x < -\tfrac{1}{2} \qquad \text{or} \qquad x > 3.$$

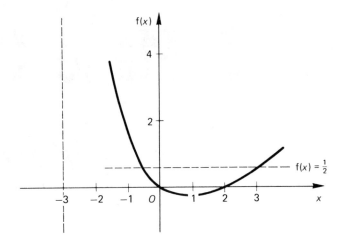

6.7

(a) $(x - y)^2 \geqslant 0$ for all real x and y,
  $\Rightarrow \quad x^2 + y^2 - 2xy \geqslant 0 \quad \Rightarrow \quad x^2 + y^2 \geqslant 2xy.$

(b) From part (a):

$$\left(\frac{a}{b}\right)^4 + \left(\frac{b}{c}\right)^4 = \left[\left(\frac{a}{b}\right)^2\right]^2 + \left[\left(\frac{b}{c}\right)^2\right]^2 \geqslant 2\left(\frac{a^2}{c^2}\right)$$

Similarly,
$$\left(\frac{c}{d}\right)^4 + \left(\frac{d}{a}\right)^4 \geqslant 2\left(\frac{c^2}{a^2}\right).$$

But
$$\left(\frac{a}{c}\right)^2 + \left(\frac{c}{a}\right)^2 \geqslant 2\left(\frac{a}{c}\right)\left(\frac{c}{a}\right) = 2,$$

hence
$$\left(\frac{a}{b}\right)^4 + \left(\frac{b}{c}\right)^4 + \left(\frac{c}{d}\right)^4 + \left(\frac{d}{a}\right)^4 \geqslant 4.$$

6.8 Let $y = \dfrac{x^2 - 2x + 1}{x^2 + 4x + 5}$.

Then $x^2(y - 1) + x(4y + 2) + (5y - 1) = 0.$ (1)
x is real, so $(4y + 2)^2 \geqslant 4(y - 1)(5y - 1)$
$\Rightarrow \quad 0 \geqslant y(y - 10)$, so $0 \leqslant y \leqslant 10.$
When $y = 0$, equation 1 gives $-x^2 + 2x - 1 = 0$
$\Rightarrow \quad x = 1$ (twice).
When $y = 10$, equation 1 gives $9x^2 + 42x + 49 = 0$
$\Rightarrow \quad x = -\tfrac{7}{3}$ (twice).
As $|x| \to \infty$, $y \to 1.$

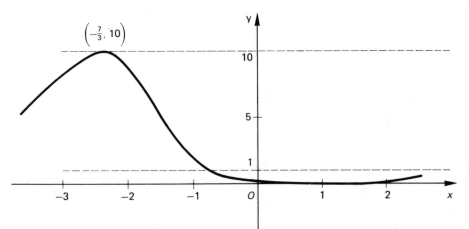

7 Plane Cartesian Coordinates

Understanding the relationship between a graph and the associated algebraic relation.

Ability to sketch curves such as $y = kx^n$ for integral and simple rational n,

$$ax + by + c = 0, \quad \frac{x^2}{a^2} + \frac{y^2}{b^2} = 1.$$

Finding the equations of straight lines and circles.

Points of intersection, distance and angle formulae.

Simple transformations $y = af(x)$, $y = f(x) + a$, $y = f(x - a)$, $y = f(ax)$.

The relation of a graph to its symmetries.

7.1 Fact Sheet

(a) Lines

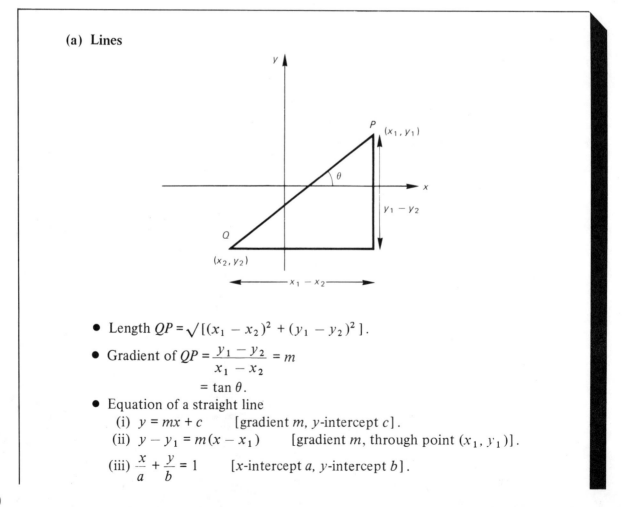

- Length $QP = \sqrt{[(x_1 - x_2)^2 + (y_1 - y_2)^2]}$.

- Gradient of $QP = \dfrac{y_1 - y_2}{x_1 - x_2} = m$

  $\qquad\qquad = \tan \theta$.

- Equation of a straight line

    (i) $y = mx + c$ [gradient m, y-intercept c].

    (ii) $y - y_1 = m(x - x_1)$ [gradient m, through point (x_1, y_1)].

    (iii) $\dfrac{x}{a} + \dfrac{y}{b} = 1$ [x-intercept a, y-intercept b].

(iv) $\dfrac{y - y_1}{x - x_1} = \dfrac{y_1 - y_2}{x_1 - x_2}$ [through (x_1, y_1) and (x_2, y_2)].

(v) $ax + by + c = 0$ [general equation].

- Distance of a point (h, k) from a line $ax + by + c = 0$ is

$$\left| \dfrac{ah + bk + c}{\sqrt{(a^2 + b^2)}} \right| \quad \text{or} \quad \pm \dfrac{ah + bk + c}{\sqrt{(a^2 + b^2)}}.$$

- Angle between two lines with gradients m_1 and m_2 is α, where

$$\tan \alpha = \dfrac{m_1 - m_2}{1 + m_1 m_2}.$$

- Lines are perpendicular if $m_1 m_2 = -1$.
- Lines are parallel if $m_1 = m_2$.

(b) Circles

(i) $(x - h)^2 + (y - k)^2 = r^2$ [centre (h, k), radius r].

(ii) $x^2 + y^2 + 2gx + 2fy + c = 0$ [centre $(-g, -f)$, radius $\sqrt{(g^2 + f^2 - c)}$].

(c) Ellipses

$$\dfrac{x^2}{a^2} + \dfrac{y^2}{b^2} = 1 \qquad \text{[semi-axes } a, b\text{]}.$$

(d) $y = kx^n$, k positive   [k negative reflects in the x-axis.]

(i) n is an even integer > 1.

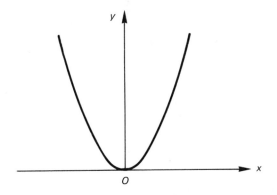

(ii) n is an odd integer > 1.

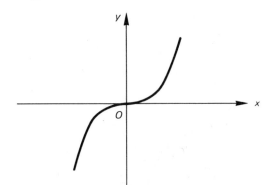

(iii) $n = \dfrac{1}{p}$, where p is a positive even integer.

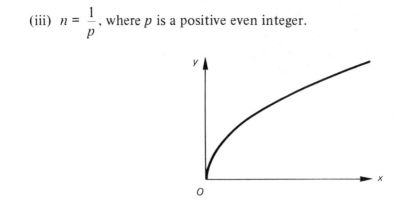

We take $y = kx^{1/2}$, for example, to mean only the positive square root so the graph does not have a branch below the x-axis.

(iv) $n = \dfrac{1}{q}$, where q is a positive odd integer.

(v) n is a negative even integer.

(vi) n is a negative odd integer.

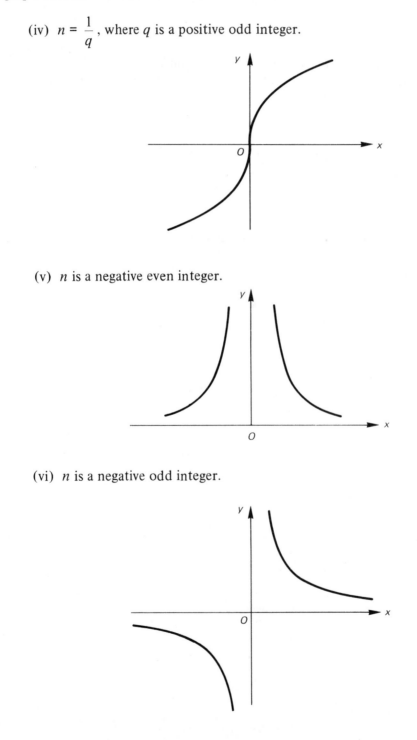

(e) Transformations

(i) $y = af(x)$ is a one-way stretch, or scaling, parallel to the y-axis, factor a.

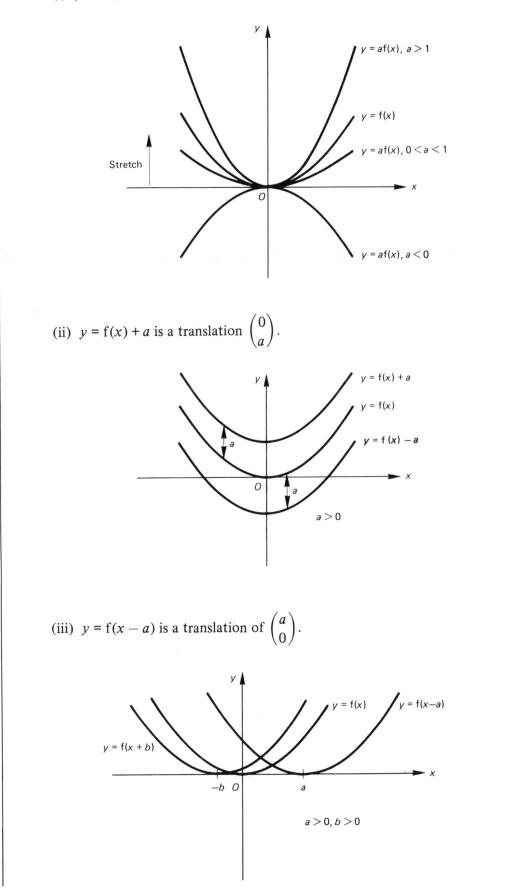

(ii) $y = f(x) + a$ is a translation $\begin{pmatrix} 0 \\ a \end{pmatrix}$.

(iii) $y = f(x - a)$ is a translation of $\begin{pmatrix} a \\ 0 \end{pmatrix}$.

(iv) $y = f(ax)$ is a one-way stretch, or scaling, parallel to the x-axis, factor $\dfrac{1}{a}$.

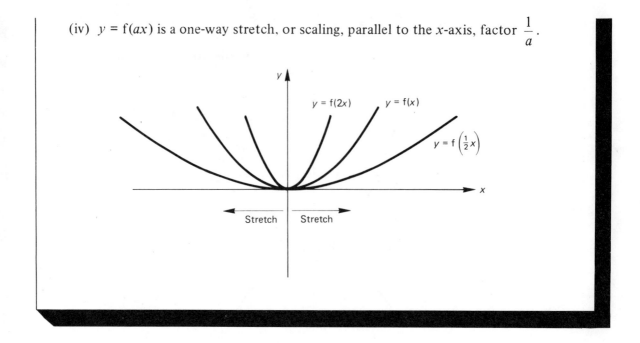

7.2 Worked Examples

7.1 Express $5x^2 - 20x + 16$ in the form $a(x - b)^2 + c$. Show how the graph of $y = 5x^2 - 20x + 16$ may be obtained from the graph of $y = x^2$ by appropriate translations and one-way stretches. List these transformations clearly in the order of application.

● $5x^2 - 20x + 16 = 5\,(x^2 - 4x + \tfrac{16}{5})$.
   Completing the square with $x^2 - 4x$,
$$x^2 - 4x + 4 = (x - 2)^2.$$
   Thus
$$x^2 - 4x + \tfrac{16}{5} = x^2 - 4x + 4 - \tfrac{4}{5} = (x - 2)^2 - \tfrac{4}{5}.$$
   Hence
$$5x^2 - 20x + 16 = 5\,[(x - 2)^2 - \tfrac{4}{5}]$$
$$= 5\,(x - 2)^2 - 4.$$
   $y = 5x^2 - 20x + 16$ may be obtained from $y = x^2$ by:

   (a) Translation $\begin{pmatrix} 2 \\ 0 \end{pmatrix}$, i.e. 2 units in the direction of the x-axis.

   (b) One-way stretch, or scaling, of factor 5 parallel to the y-axis.

   (c) Translation $\begin{pmatrix} 0 \\ -4 \end{pmatrix}$, i.e. -4 units in the direction of the y-axis.

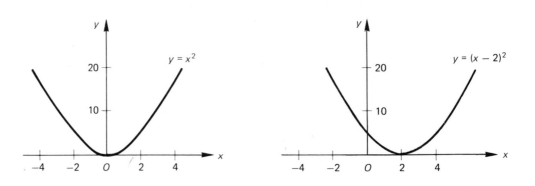

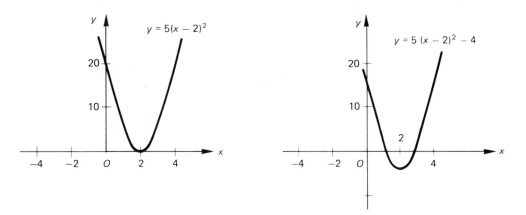

7.2 The circle with equation $x^2 + y^2 + 8x + 6y - 56 = 0$ has radius

A, 5; B, 9; C, $\sqrt{56}$; D, $\sqrt{103}$; E, $\sqrt{156}$.

- The equation of a circle is $(x - a)^2 + (y - b)^2 = r^2$, where (a, b) is the centre and r is the radius.
$$x^2 + y^2 + 8x + 6y - 56 = 0 \quad \Rightarrow \quad x^2 + 8x + y^2 + 6y = 56.$$
$x^2 + 8x$ requires $(\frac{8}{2})^2 = 16$ to complete the square,
$y^2 + 6y$ requires $(\frac{6}{2})^2 = 9$ to complete the square.
Add $16 + 9$ to each side of the equation of the circle:
$$x^2 + 8x + 16 + y^2 + 6y + 9 = 56 + 16 + 9 = 81$$
$$\Rightarrow \quad (x + 4)^2 + (y + 3)^2 = 81 \quad \Rightarrow \quad \text{centre } (-4, -3), \text{ radius } 9.$$

Answer **B**

7.3 If the points (h, k), $(1, 3)$ and $(-2, 7)$ are collinear, then the relationship connecting h and k could be:

A, $3h + 4k = 5$; B, $3k - 4h = 5$; C, $3k + 4h = 13$;
D, $3h - 4k = 13$; E, $3k + 4h = -13$.

- If the points P (h, k), A $(1, 3)$ and B $(-2, 7)$ are collinear, then PA has the same gradient as AB.

Gradient of PA is $\dfrac{k - 3}{h - 1}$, gradient of AB is $\dfrac{7 - 3}{-2 - 1} = -\dfrac{4}{3}$.

Thus $\qquad \dfrac{k - 3}{h - 1} = -\dfrac{4}{3} \quad \Rightarrow \quad 3(k - 3) = -4(h - 1)$

$$3k - 9 = -4h + 4 \quad \Rightarrow \quad 3k + 4h = 13. \qquad \text{Answer } \mathbf{C}$$

7.4 The circle which passes through the origin and the points $(25, 0)$ and $(16, 12)$ has the equation

A, $(x - 10)^2 + (y - 6)^2 = 16$; B, $x^2 + y^2 - 20x + 12y = 0$;
C, $x^2 + y^2 - 25x - 12y = 0$; D, $(x - 25)^2 + y^2 = 625$;
E, $x^2 + y^2 - 25x = 0$.

- The general equation of a circle is $x^2 + y^2 + 2gx + 2fy + c = 0$.
Since the circle passes through $(0, 0)$ then $c = 0$.
Since the circle passes through $(25, 0)$ then
$625 + 50g = 0 \quad \Rightarrow \quad g = -\frac{25}{2}$.
If the circle passes through $(16, 12)$ then
$256 + 144 + 32g + 24f = 0 \quad \Rightarrow \quad 400 - 400 + 24f = 0$, so $f = 0$.
Hence equation of the circle is $x^2 + y^2 - 25x = 0$.

Answer **E**

7.5   Find the centre and radius of the circle
$$x^2 + y^2 - 6x - 10y + 9 = 0.$$
Find the points of intersection of the line $y = 2x + 4$ and the given circle, and prove that the length of the chord cut off is $4\sqrt{5}$. Show that the circle which has the same centre as the given circle and which touches the given line passes through the point $(1, 4)$.

What is the equation of the tangent to the second circle at $(1, 4)$?

● $x^2 + y^2 - 6x - 10y + 9 = 0 \Rightarrow x^2 - 6x + y^2 - 10y = -9$
  $\Rightarrow (x - 3)^2 + (y - 5)^2 = -9 + 9 + 25 = 25 \equiv S_1$.
Circle: centre $(3, 5)$, radius 5.
When $y = 2x + 4$ intersects $x^2 + y^2 - 6x - 10y + 9 = 0$,
$$x^2 + (2x + 4)^2 - 6x - 10(2x + 4) + 9 = 0$$
$$\Rightarrow x^2 + 4x^2 + 16x + 16 - 6x - 20x - 40 + 9 = 0$$
$$\Rightarrow 5x^2 - 10x - 15 = 0 \Rightarrow 5(x + 1)(x - 3) = 0,$$
so $x = -1$ or $x = 3$.

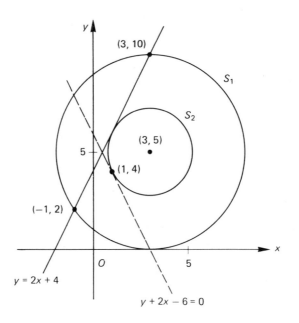

Substituting for x into $y = 2x + 4$ gives intersection points $(-1, 2)$ and $(3, 10)$.
Length of chord $= \sqrt{[(3 - -1)^2 + (10 - 2)^2]} = 4\sqrt{5}$.
Any circle with centre $(3, 5)$ and radius r has equation:
$$(x - 3)^2 + (y - 5)^2 = r^2.$$
Distance of $(3, 5)$ from the line $y = 2x + 4$ is
$$\left| \frac{(2)(3) - 5 + 4}{\sqrt{(2^2 + 1^2)}} \right| = \sqrt{5}.$$

Hence the equation of the circle which touches the line $y = 2x + 4$ is
$(x - 3)^2 + (y - 5)^2 = 5 \equiv S_2$

When $y = 4$, $x^2 - 6x + 9 + 1 = 5$

$\Rightarrow x^2 - 6x + 5 = 0$

$\Rightarrow (x - 5)(x - 1) = 0$,

so $x = 5$ or $x = 1$.

Hence the circle S_2 passes through $(1, 4)$.

The gradient of the radius of S_2 which passes through $(1, 4)$ is $\dfrac{5 - 4}{3 - 1} = \dfrac{1}{2}$,

therefore the gradient of the tangent at $(1, 4)$ is -2.

The equation of the tangent is

$$y - 4 = -2(x - 1) \quad \Rightarrow \quad y + 2x - 6 = 0.$$

7.6   Sketch on the same diagram the curves whose equations are $x^2 + y^2 = 25$ and $x^2 + 4y = 0$. (Do not calculate the coordinates of the points of inter-section.) Shade in your diagram the regions of the plane for which

$$(x^2 + y^2 - 25)(x^2 + 4y) < 0.$$

● $x^2 + y^2 = 25$ is the equation of a circle centre $(0, 0)$ radius 5.

$x^2 + 4y = 0$ is the equation of a parabola $y = -\frac{1}{4}x^2$.

$x^2 + y^2 - 25 < 0$ for points inside the circle (shaded horizontally in the left-hand diagram).

$x^2 + 4y < 0$ for points below the parabola (shaded vertically in the same diagram).

$(x^2 + y^2 - 25)(x^2 + 4y) < 0$ when one factor is positive and one is nega-tive (shaded horizontally in the right-hand diagram).

(Boundaries are not included.)

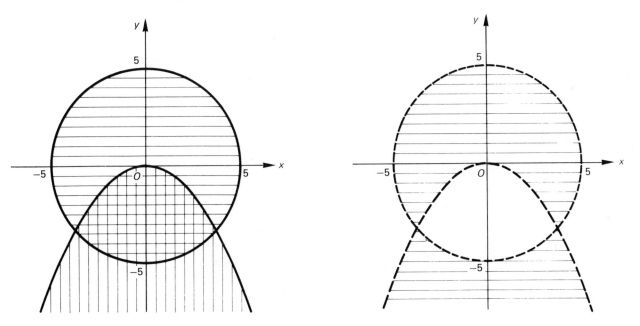

7.7   A function g(x) of period 2π is defined by

$$g(x) = x^2 \qquad \text{for} \qquad 0 \leqslant x \leqslant \frac{\pi}{2},$$

$$g(x) = \frac{\pi^2}{4} \qquad \text{for} \qquad \frac{\pi}{2} < x \leqslant \pi.$$

Given also that $g(x) = g(-x)$ for all x, sketch the graph of $g(x)$ for $-2\pi \leqslant x \leqslant 2\pi$.

(L)

● $g(x) = x^2$ for $0 \leqslant x \leqslant \dfrac{\pi}{2}$ gives figure (a).

$g(x) = \dfrac{\pi^2}{4}$ for $\dfrac{\pi}{2} < x \leqslant \pi$ gives figure (b).

Combining these two gives figure (c).
$g(x) = g(-x)$ extends the graph to figure (d).
$g(x)$ has a period of 2π, so this part of the curve is one period. Extending it to $-2\pi \leqslant x \leqslant 2\pi$ gives figure (e).

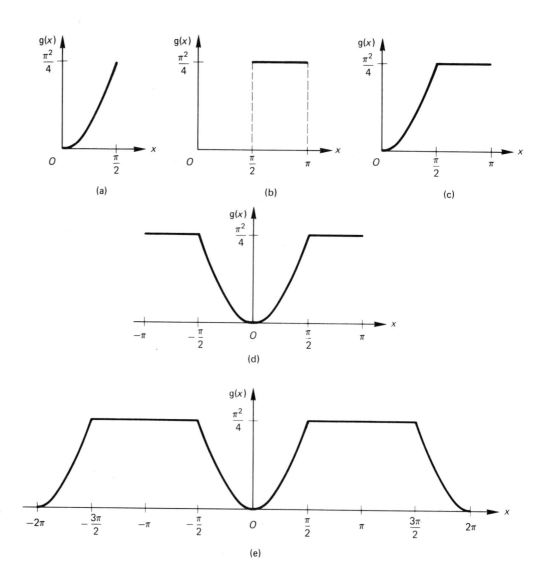

7.8 The curve whose equation is $y = x^2$ undergoes, in succession, the following transformations:
(a) A translation of magnitude 3 units in the direction of the negative x-axis.
(b) A scaling parallel to the y-axis by a factor of 2.

(c) A translation of magnitude 3 units in the direction of the y-axis.

Give the equation of the resulting curve and sketch this curve. (Your sketch should show the coordinates of at least three points on the curve.)

Another curve undergoes, in succession, the transformations (a), (b), (c) as above, and the equation of the resulting curve is $y = \dfrac{5x + 12}{x + 2}$.

Determine the equation of the curve before the three transformations were effected.

Sketch the curve with the equation $y = \dfrac{5x + 12}{x + 2}$.

- $y = x^2 \;\Rightarrow\; y = (x + 3)^2$ after translation $\begin{pmatrix} -3 \\ 0 \end{pmatrix}$.

 $y = (x + 3)^2 \;\Rightarrow\; y = 2(x + 3)^2$ after scaling factor of 2.

 $y = 2(x + 3)^2 \;\Rightarrow\; y = 2(x + 3)^2 + 3$ after translation $\begin{pmatrix} 0 \\ 3 \end{pmatrix}$.

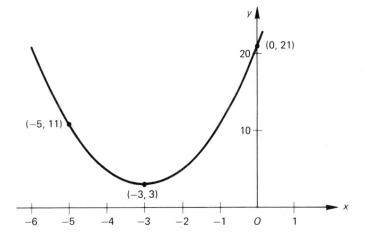

$y = \dfrac{5x + 12}{x + 2}$ has to undergo transformations $(c)^{-1}$, $(b)^{-1}$ and $(a)^{-1}$ in succession in order to find the equation before the transformations (a), (b) and (c):

$$y = \frac{5x + 12}{x + 2} \;\Rightarrow\; y = \frac{5x + 12}{x + 2} - 3 = \frac{2x + 6}{x + 2}. \qquad \text{under } (c)^{-1}$$

$$y = \frac{2x + 6}{x + 2} \;\Rightarrow\; y = \frac{1}{2}\left(\frac{2x + 6}{x + 2}\right) = \frac{x + 3}{x + 2}. \qquad \text{under } (b)^{-1}$$

$$y = \frac{x + 3}{x + 2} \;\Rightarrow\; y = \frac{(x - 3) + 3}{(x - 3) + 2} = \frac{x}{x - 1}. \qquad \text{under } (a)^{-1}$$

See diagram on next page.

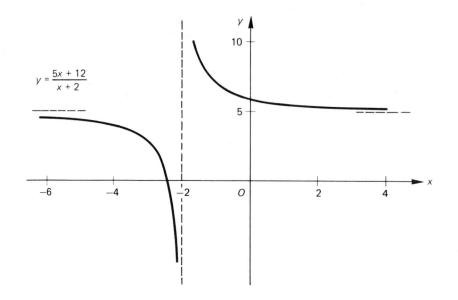

$$y = \frac{5x + 12}{x + 2}$$

7.9 The vertex A of a square $ABCD$ is at the point $(4, -3)$. The diagonal BD has equation $x - 7y + 75 = 0$, and the vertex D is nearer to the origin than B. Calculate

(a) the coordinates of the centre of the square and of B, C and D,

(b) the length of AB,

(c) the equation of the circle which touches the sides of the square.

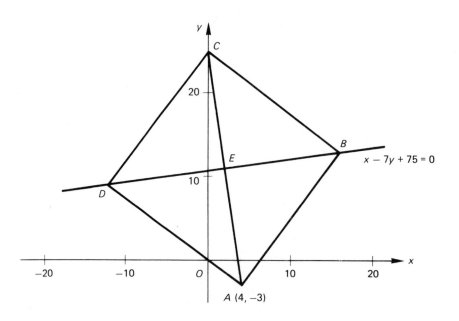

- (a) Line BD is $x - 7y + 75 = 0$ $\Rightarrow$ $y = \frac{1}{7}x + \frac{75}{7}$.   (1)

  Gradient of BD is $\frac{1}{7}$.

  The diagonals of a square are perpendicular, so AC has a gradient of -7.

  Equation of AC is $y - (-3) = -7(x - 4)$,

  $\Rightarrow$ $y + 3 = -7x + 28$ or $7x + y = 25$ (2)

  AC and BD intersect at the centre of the square.

  Solving equations 1 and 2 simultaneously,

  $x - 7(25 - 7x) + 75 = 0$ $\Rightarrow$ $50x = 100,$ $x = 2, y = 11.$

  Therefore the centre of the square E is the point $(2, 11)$.

A is $(4, -3)$, E is $(2, 11)$. Therefore $\overline{EA} = \begin{pmatrix} 2 \\ -14 \end{pmatrix}$. Since the diagonals of a square are equal and bisect perpendicularly at E,

$$\overline{EB} = \begin{pmatrix} 14 \\ 2 \end{pmatrix}, \quad \overline{EC} = \begin{pmatrix} -2 \\ 14 \end{pmatrix}, \quad \overline{ED} = \begin{pmatrix} -14 \\ -2 \end{pmatrix}.$$

Thus B is $(2 + 14, 11 + 2)$, C is $(2 - 2, 11 + 14)$, D is $(2 - 14, 11 - 2)$.

$$\Rightarrow \quad B\,(16, 13), \quad C\,(0, 25), \quad D\,(-12, 9).$$

(D is nearer to O than B.)

(b) The length of AB is $\sqrt{[(16 - 4)^2 + (13 + 3)^2]} = 20$ units.

(c) If a circle touches all the sides of the square, the centre is at $E\,(2, 11)$ and the diameter is 20 units, so the radius is 10 units.

The equation of the circle is
$$(x - 2)^2 + (y - 11)^2 = 10^2$$
$$\Rightarrow \quad x^2 + y^2 - 4x - 22y + 25 = 0.$$

7.3 Exercises

7.1 The equation of the line through the points $(1, -2)$ and $(-5, 6)$ is

A, $4x - 3y = -1$;   B, $4x + 3y = -2$;   C, $3x + 4y = -5$;
D, $3x - 4y = 11$;   E, none of these.

7.2 The locus of the points equidistant from the centres of the circles whose equations are
$$x^2 + y^2 - 2x + 4y - 5 = 0$$
and
$$x^2 + y^2 - 10x - 8y + 2 = 0 \qquad\qquad \text{has equation}$$

A, $8x + 12y - 7 = 0$;   B, $6x + 2y + 3 = 0$;   C, $4x + 6y - 5 = 0$;
D, $2x + 3y - 9 = 0$;   E, none of these.

7.3 The centre of the circle $3x^2 + 3y^2 - 15x - 6y + 2 = 0$ is the point

A, $(15, 6)$;   B, $(7.5, 3)$;   C, $(-5, -2)$;   D, $(-15, -6)$;   E, $(2.5, 1)$.

7.4 A and B are the points $(2, 4)$ and $(4, 10)$ respectively. Find
(a) the equation of AB,
(b) the equation of the perpendicular bisector of AB,
(c) the equations of circles through A which touch both of the coordinate axes,
(d) the perpendicular distance of the centres of the circles from the perpendicular bisector of AB,
(e) the point other than A at which the line AB intersects the smaller circle.

7.5 The curve whose equation is $y = x^3$ undergoes in succession the following transformations:
(a) a translation of magnitude 2 parallel to the y-axis,
(b) a scaling parallel to the x-axis of factor 2,

(c) a scaling parallel to the y-axis of factor 3,
(d) a translation of -2 parallel to the y-axis.

   Give the equation and sketch of the resulting curve, and the coordinates of three points on the curve.

7.6 Find the equation of the circle having the line joining the points $(2, 1)$ and $(4, 7)$ as diameter. Find the equations of the tangents from the point $(3, 0)$ to this circle.

7.7 The coordinates of the vertices of the triangle ABC are $A(2, 0)$, $B(8, -3)$, $C(5, 6)$. Show that the triangle is isosceles and find the midpoint D of BC. Find the equation of the circumcircle of triangle ACD. Find the angle ABD (a construction will not be accepted).

7.4 Outline Solutions to Exercises

7.1 $(1, -2)$ does not satisfy A, but does satisfy B, C and D.
$(-5, 6)$ satisfies B, but not C or D <u>Answer B</u>

7.2 Centres of circles are C_1 $(1, -2)$ and C_2 $(5, 4)$. $P\,(x, y)$ is equidistant from C_1 and C_2 if $(PC_1)^2 = (PC_2)^2$.
$\Rightarrow \quad (x - 1)^2 + (y + 2)^2 = (x - 5)^2 + (y - 4)^2$
$\Rightarrow \quad 8x + 12y = 36 \quad \Rightarrow \quad 2x + 3y = 9.$ <u>Answer D</u>

7.3 $x^2 + y^2 - 5x - 2y + \frac{2}{3} = 0$ has centre $(\frac{5}{2}, \frac{2}{2})$.   <u>Answer E</u>

7.4

(a) $\dfrac{y - 4}{4 - 10} = \dfrac{x - 2}{2 - 4} \quad \Rightarrow \quad y = 3x - 2.$

(b) Midpoint of AB is $(3, 7)$, perpendicular bisector has gradient $-\frac{1}{3}$, so equation is
$$y - 7 = -\tfrac{1}{3}(x - 3) \quad \Rightarrow \quad 3y = -x + 24.$$

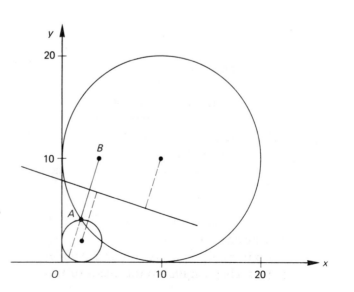

(c) If circle has centre (h, k), in order to touch both axes $h = \pm k$ and radius $= |h|$. Equation has form $(x - h)^2 + (y \pm h)^2 = h^2$.

To go through $(2, 4)$, circle lies in first quadrant, so $h = +k$ and
$$(2 - h)^2 + (4 - h)^2 = h^2 \quad \Rightarrow \quad h = 10 \quad \text{or} \quad h = 2.$$
Circles are $(x - 10)^2 + (y - 10)^2 = 100$
and $\qquad (x - 2)^2 + (y - 2)^2 = 4$.

(d) Centres are $(10, 10)$ and $(2, 2)$.

Perpendicular distance of $(10, 10)$ from $x + 3y - 24 = 0$ is
$$\left| \frac{10 + (3)(10) - 24}{\sqrt{(1^2 + 3^2)}} \right| = \frac{16}{\sqrt{10}}.$$

Perpendicular distance of $(2, 2)$ from $x + 3y - 24 = 0$ is
$$\left| \frac{2 + (3)(2) - 24}{\sqrt{(1^2 + 3^2)}} \right| = \frac{16}{\sqrt{10}}.$$

(e) AB has equation $y = 3x - 2$, smaller circle has equation $(x - 2)^2 + (y - 2)^2 = 4$. Substituting gives
$$(x - 2)^2 + (3x - 4)^2 = 4 \quad \Rightarrow \quad x = 2 \quad \text{or} \quad x = \tfrac{4}{5}.$$
Required point is $(\tfrac{4}{5}, \tfrac{2}{5})$.

7.5 $\quad y = x^3$

$y_A = x^3 + 2.$

$y_B = \left(\dfrac{x}{2}\right)^3 + 2.$

$y_C = 3\left[\left(\dfrac{x}{2}\right)^3 + 2\right].$

$y_D = 3\left[\left(\dfrac{x}{2}\right)^3 + 2\right] - 2 \quad \Rightarrow \quad 8y = 3x^3 + 32.$

Three points $(0, 4)$, $(2, 7)$, $(-2, 1)$.

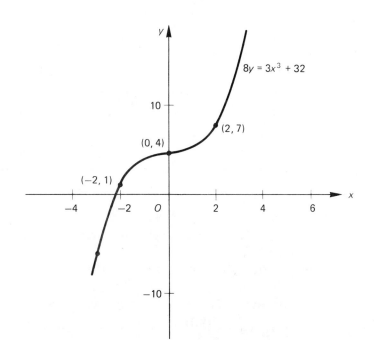

7.6 Centre of circle is $(3, 4)$,
$$\text{radius} = \sqrt{(1 + 9)} = \sqrt{(10)} \quad \Rightarrow \quad (x - 3)^2 + (y - 4)^2 = 10.$$
Line through $(3, 0)$ is $y = m(x - 3)$.
To be a tangent, perpendicular distance from $(3, 4) = \sqrt{(10)}$.

$$\frac{m(3) - 4 - 3m}{\sqrt{(m^2 + 1^2)}} = \pm\sqrt{(10)} \quad \Rightarrow \quad 10m^2 = 6 \quad \Rightarrow \quad m = \pm\sqrt{(\tfrac{3}{5})}.$$

Equations of tangents are $\quad y = \pm\sqrt{(\tfrac{3}{5})}\,(x - 3)$.

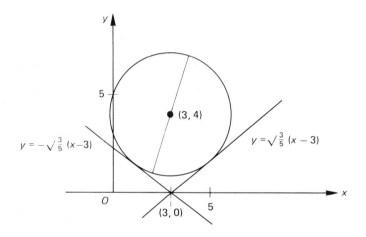

7.7 $AB = \sqrt{(36 + 9)}$, $BC = \sqrt{(9 + 81)}$, $CA = \sqrt{(9 + 36)}$.
$AB = CA \quad \Rightarrow \quad$ isosceles triangle.

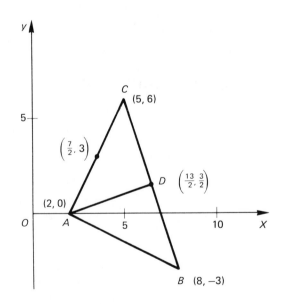

Midpoint D of BC is $(\tfrac{13}{2}, \tfrac{3}{2})$.
Since $\angle ADC$ is a right angle, AC must be a diameter of the circumcircle through A, D and C,
$\Rightarrow \quad$ centre is $(\tfrac{7}{2}, 3)$, radius is $\tfrac{1}{2}\sqrt{(45)}$,
equation $\quad (x - \tfrac{7}{2})^2 + (y - 3)^2 = \tfrac{45}{4}$,
$\Rightarrow \quad x^2 + y^2 - 7x - 6y + 10 = 0$.

$$\cos\angle ABD = \frac{\sqrt{(90)}}{2\sqrt{(45)}} \quad \Rightarrow \quad \angle ABD = 45°.$$

8 Trigonometric Functions and Formulae

Definition of the six trigonometric functions for any angle; their periodic properties and symmetries.

Use of the sine and cosine formulae.

The angle between a line and a plane, between two planes and between two skew lines, using trigonometric methods.

Circular measure.

8.1 Fact Sheet

(a) Degrees and Radians

- 1 radian is defined as the angle subtended at the centre of a circle by an arc equal in length to the radius.

- 1 radian = $\dfrac{180}{\pi}$ degrees.

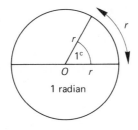

1 radian

(b) Sine, Cosine and Tangent of Any Angle

- $\sin \theta = \dfrac{\text{projection of } OP \text{ onto } OY}{\text{length of } OP}$.

- $\cos \theta = \dfrac{\text{projection of } OP \text{ onto } OX}{\text{length of } OP}$.

- $\tan \theta = \dfrac{\sin \theta}{\cos \theta}$.

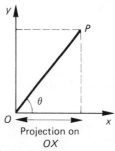

Projection on OX

- In the second quadrant:
  $\sin \theta = \sin(180 - \theta)$,
  $\cos \theta = -\cos(180 - \theta)$,
  $\tan \theta = -\tan(180 - \theta)$.
- In the third quadrant:
  $\sin \theta = -\sin(\theta - 180)$,
  $\cos \theta = -\cos(\theta - 180)$,
  $\tan \theta = \tan(\theta - 180)$.
- In the fourth quadrant:
  $\sin \theta = -\sin(360 - \theta)$,
  $\cos \theta = \cos(360 - \theta)$,
  $\tan \theta = -\tan(360 - \theta)$.

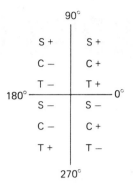

(c) Cosecant, Secant and Cotangent

- $\operatorname{cosec} \theta = \dfrac{1}{\sin \theta}$.

- $\sec \theta = \dfrac{1}{\cos \theta}$.

- $\cot \theta = \dfrac{1}{\tan \theta} = \dfrac{\cos \theta}{\sin \theta}$.

(d) Pythagorean Identities

- $\cos^2 \theta + \sin^2 \theta = 1$.
- $1 + \tan^2 \theta = \sec^2 \theta$.
- $\cot^2 \theta + 1 = \operatorname{cosec}^2 \theta$.

(e) Sine and Cosine Formulae

- Sine
  $$\frac{a}{\sin A} = \frac{b}{\sin B} = \frac{c}{\sin C} = 2R \text{ where } R \text{ is the radius of the circumcircle.}$$
- Cosine
  $$a^2 = b^2 + c^2 - 2(b)(c)\cos A \quad \text{or} \quad \cos A = \frac{b^2 + c^2 - a^2}{2(b)(c)}.$$
  In the special case when angle A is a right angle, $\sin A = 1$, $\cos A = 0$, and they

  become $\qquad a = \dfrac{b}{\sin B} = \dfrac{c}{\sin C} \; ; \qquad a^2 = b^2 + c^2.$

(f) Circular Measure

- Length of an arc = $r\theta$, and area of a sector = $\frac{1}{2} r^2 \theta$, where θ is measured in radians.

(g) Graphs of the Six Trigonometric Functions

● sin θ

Period 360° or 2π radians.
$\sin\theta = \sin[180n° + (-1)^n\,\theta]$.

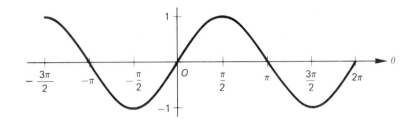

● cos θ

Period 360° or 2π radians.
$\cos\theta = \cos(360n° \pm \theta)$.

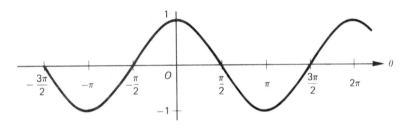

● tan θ

Period 180° or π radians.
$\tan\theta = \tan(180n° + \theta)$.

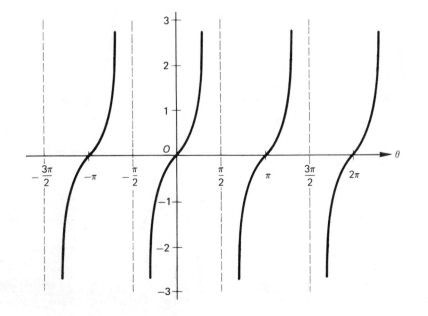

● cosec θ Behaves as sin θ.

● sec θ Behaves as cos θ.

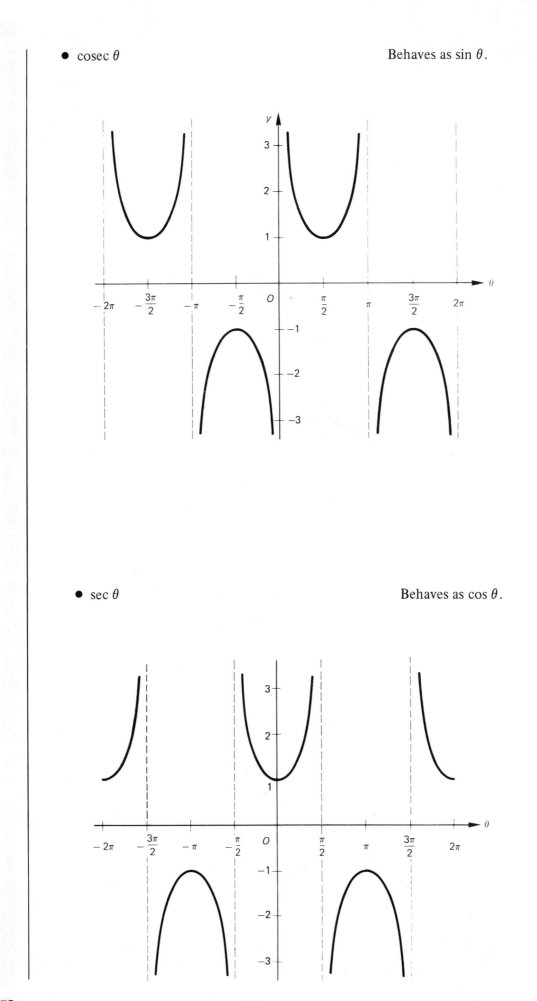

- cotan θ Behaves as tan θ.

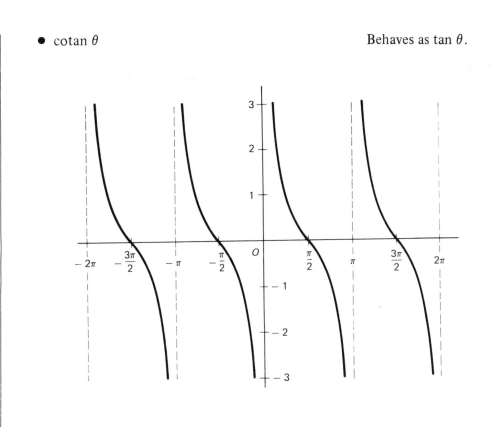

(h) Skew Lines

- Lines which are not parallel but do not intersect are called skew lines. Examples:
  (i) Edges AB and $A'D'$ of the cube shown.

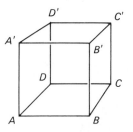

(ii) Edges VA and CB of the pyramid shown.

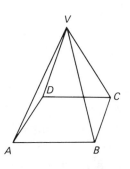

8.2 Worked Examples

8.1 *ABC* is an equilateral triangle of side 4 cm. The radius, in cm, of the circle passing through *A*, *B* and *C* is

A, 4;   B, 2;   C, $2\sqrt{3}$;   D, $4/\sqrt{3}$;   E, $\sqrt{3}/2$.

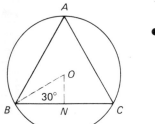

● Let *O* be the centre of the circle, the point where the perpendicular bi-sectors of the sides intersect.
OB bisects $\angle ABC \Rightarrow \angle OBC = 30°$.
Let *N* be the mid-point of *BC*.
$BN = 2$ cm, $\angle ONB = 90°$.
In triangle *OBN*,

$$\frac{BN}{OB} = \cos 30°, \text{ i.e. } \frac{2}{OB} = \frac{\sqrt{3}}{2} \Rightarrow OB = \frac{4}{\sqrt{3}}.$$ Answer **D**

Alternatively: by the sine rule $2R = \dfrac{a}{\sin A}$, $R = \dfrac{2}{\sin 60°} = \dfrac{4}{\sqrt{3}}$.

8.2 With the usual notation in a triangle *XYZ*, $x = 7$, $y = 4$, $z = 5$. Without using tables, find $\cos X$ as a fraction in its lowest terms and prove that

$$\sin Y = \frac{8\sqrt{6}}{35}.$$

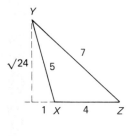

● Using the cosine rule for triangle *XYZ*,

$$\cos X = \frac{y^2 + z^2 - x^2}{2yz} = \frac{16 + 25 - 49}{(2)(4)(5)} = -\tfrac{1}{5}.$$

If $\cos X = -\tfrac{1}{5}$ and $\sin X^2 + \cos^2 X = 1$
then

$$\sin^2 X = 1 - (\tfrac{1}{5})^2 = \frac{24}{25} \Rightarrow \sin X = \frac{\sqrt{24}}{5} = \frac{2\sqrt{6}}{5}$$

(positive value only since $0° < X < 180°$).
Using the sine rule for triangle *XYZ*,

$$\frac{\sin Y}{y} = \frac{\sin X}{x} \Rightarrow \sin Y = \frac{y \sin X}{x} = \frac{(4)(2\sqrt{6})}{(7)(5)} = \frac{8\sqrt{6}}{35}.$$

8.3 In triangle *ABC*, angle $A = \dfrac{\pi}{3}$.

(a) Prove that $\sin C = \tfrac{1}{2}(\sqrt{3} \cos B + \sin B)$.
(b) Given that $c = 3b$, where the usual notation applies, find, by using the sine rule or otherwise, the size of angle *B*.

● (a) $A + B + C = \pi$ (angle sum of a triangle).
  $\Rightarrow C = \pi - (A + B)$, $\sin C = \sin [\pi - (A + B)] = \sin (A + B)$
  $\Rightarrow \sin C = \sin A \cos B + \cos A \sin B = \sin \dfrac{\pi}{3} \cos B + \cos \dfrac{\pi}{3} \sin B$

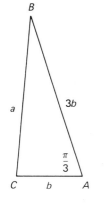

$$\sin C = \frac{\sqrt{3}}{2} \cos B + \frac{1}{2} \sin B = \frac{1}{2} (\sqrt{3} \cos B + \sin B). \qquad (1)$$

(b) By the sine rule, $\dfrac{\sin B}{b} = \dfrac{\sin C}{c}$, i.e. $\sin C = \dfrac{c \sin B}{b}$.

But $c = 3b$, therefore $\sin C = 3 \sin B$. $\qquad (2)$
Combining equations 1 and 2,
$$3 \sin B = \tfrac{1}{2}(\sqrt{3} \cos B + \sin B),$$
$$5 \sin B = \sqrt{3} \cos B.$$
Divide by $5 \cos B$; $\tan B = \dfrac{\sqrt{3}}{5} \quad \Rightarrow \quad B = 0.333$ radians.

8.4 The length of an edge of a regular tetrahedron $ABCD$ is $3a$. Show that the perpendicular distance of the vertex from the opposite face is $a\sqrt{6}$.

A point P lies on the edge AB and α is the acute angle which DP makes with the plane ABC. Prove that as P moves on the edge AB the greatest and least values of α are $70.5°$ and $54.7°$ approximately.

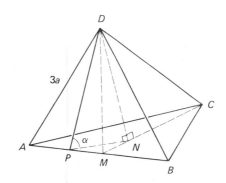

Let N be the foot of the perpendicular from D to face ABC and M the midpoint of AB.
ABC is an equilateral triangle $\Rightarrow CMA = 90°$.
By the theorem of Pythagoras, for triangle CMA, $CM^2 + AM^2 = AC^2$,

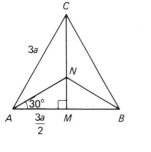

i.e. $CM^2 = 9a^2 - \dfrac{9a^2}{4} = \dfrac{27a^2}{4}$,

$$\Rightarrow \quad CM = \frac{3\sqrt{3}}{2} a = DM \text{ (symmetry)}.$$

In triangle AMN, $\dfrac{NM}{AM} = \tan 30°$, $\quad NM = \left(\dfrac{1}{\sqrt{3}}\right)\left(\dfrac{3a}{2}\right) = \dfrac{\sqrt{3}a}{2}$.

From triangle DNM, $DN^2 + NM^2 = DM^2$,

$$DN^2 = \frac{27a^2}{4} - \frac{3a^2}{4} = \frac{24a^2}{4},$$

$$DN = a\sqrt{6}.$$

The perpendicular distance of the vertex from the opposite face is $a\sqrt{6}$.

Let P be distance x from M.

Then, from triangle PMN, $PN^2 = NM^2 + PM^2 = \dfrac{3a^2}{4} + x^2$.

The angle α between DP and plane ABC is angle DPN,

$$\cot \alpha = \frac{PN}{DN} = \frac{\sqrt{(\frac{3}{4}a^2 + x^2)}}{a\sqrt{6}} \ .$$

Maximum value of $\cot \alpha$ is given by the maximum value of x, $\frac{3a}{2}$.

Minimum value of $\cot \alpha$ is given by the minimum value of x, 0.

For min. α, $\cot \alpha = \frac{\sqrt{(12a^2/4)}}{a\sqrt{6}} = \frac{1}{\sqrt{2}}$, $\alpha = 54.7°$.

For max. α, $\cot \alpha = \frac{\sqrt{(3a^2/4)}}{a\sqrt{6}} = \frac{1}{2\sqrt{2}}$, $\alpha = 70.5°$.

8.5 ABC is a plane triangle with $AB = 4$ cm, $AC = 5$ cm and angle $ACB = 30°$. Find, by calculation, the two possible lengths of BC, each correct to 3 significant figures, and the corresponding values of angle ABC, correct to the nearest tenth of a degree.

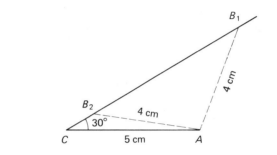

Using the cosine rule for triangle ABC,
$BA^2 = BC^2 + AC^2 - 2(BC)(AC) \cos C$,
$16 = BC^2 + 25 - 2(BC)(5) \cos 30°$,
$\Rightarrow \quad BC^2 - 8.66\, BC + 9 = 0$

$\Rightarrow \quad BC = \dfrac{8.66 \pm \sqrt{(75 - 36)}}{2} = \dfrac{8.66 \pm \sqrt{39}}{2}$

$\Rightarrow \quad BC = 7.45$ or 1.21 cm.

Using the sine rule for triangle ABC, $\dfrac{\sin \angle ABC}{AC} = \dfrac{\sin \angle BCA}{AB}$,

$\Rightarrow \quad \sin \angle ABC = \dfrac{5 \sin 30°}{4} = 0.625$

$\angle ABC = 38.7°$ or $141.3°$.
When $BC = 7.45$ cm, $\angle ABC = 38.7°$; when $BC = 1.21$ cm, $\angle ABC = 141.3°$.

8.6 A cubical packing case of edge 1 m has base $ABCD$ resting on the horizontal ground, with edges AA', BB', CC' and DD' vertical. In order to slide the case along one edge, the case is rotated about edge AB through an angle θ where $0° < \theta < 45°$.
(a) Find, in terms of θ, the heights of C and C' above the ground.
(b) Show that the diagonal BD of the base of the packing case is inclined to the horizontal at an angle ψ, where $\sin \psi = \dfrac{\sin \theta}{\sqrt{2}}$.

(c) Show that the diagonal AC' of the case is inclined to the horizontal at an angle ϕ, where $\sin \phi = \dfrac{\sin \theta + \cos \theta}{\sqrt{3}}$.

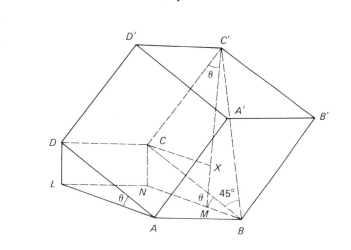

Since all faces are squares of side 1 m, each diagonal of a face is of length $\sqrt{2}$ m (theorem of Pythagoras).

(a) Let the points N and M be points on the ground vertically below C and C', $\angle CBN = \theta°$, $CB = 1$ m.

From triangle CNB, $CN = CB \sin \theta = \sin \theta$.

From triangle $CC'X$, $C'X = CC' \cos \theta = \cos \theta$.

But $C'M = C'X + XM = C'X + CN = \cos \theta + \sin \theta$.

Therefore C and C' are $\sin \theta$ m and $(\sin \theta + \cos \theta)$ m above the ground respectively.

(b) D is distance $\sin \theta$ above the ground, $DB = \sqrt{2}$ m.

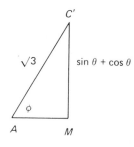

From triangle DBL, $\dfrac{DL}{DB} = \sin \psi \implies \dfrac{\sin \theta}{\sqrt{2}} = \sin \psi$.

(c) The diagonals of the packing case are each of length $\sqrt{(1^2 + 1^2 + 1^2)} = \sqrt{3}$ m.

From triangle $C'MA$, $\sin \phi = \dfrac{C'M}{C'A} = \dfrac{\sin \theta + \cos \theta}{\sqrt{3}}$.

8.7 $ABCD$ is one face of a cube and AA', BB', CC' and DD' are edges, each of length $2a$. The midpoint of AB is X, and the midpoint of DD' is Y. Calculate
(a) the lengths $B'X$, $B'Y$, XY,
(b) $\sin \angle XYB'$,
(c) the area of triangle XYB'.

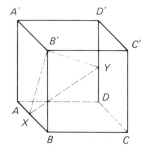

● (a) In triangle $BB'X$, $BB' = 2a$, $BX = a$, $\angle B'BX = 90°$.

By the theorem of Pythagoras, $B'X^2 = BX^2 + BB'^2$
$$B'X^2 = 5a^2, \quad B'X = \sqrt{5}a.$$

Similarly, $DX = \sqrt{5}a$.

In triangle XDY, $XD = \sqrt{5}a$, $DY = a$, $\angle XDY = 90°$.

By the theorem of Pythagoras, $XY^2 = XD^2 + DY^2$
$$XY^2 = 6a^2, \quad XY = \sqrt{6}a.$$

In triangle $A'B'D'$, $A'D' = A'B' = 2a$, $\angle A'B'D' = 45°$.

83

$$\frac{B'A'}{B'D'} = \cos 45° = \frac{1}{\sqrt{2}} \quad \Rightarrow \quad B'D' = 2\sqrt{2}a.$$

In triangle $B'D'Y$, $B'D' = 2\sqrt{2}a$, $D'Y = a$, $\angle B'D'Y = 90°$.
By the theorem of Pythagoras, $B'Y^2 = B'D'^2 + D'Y^2$
$$B'Y^2 = 9a^2, \qquad B'Y = 3a.$$
$$B'X = \sqrt{5}a, \qquad B'Y = 3a, \qquad XY = \sqrt{6}a.$$

(b) By the cosine rule for triangle $B'XY$,
$$\cos \angle XYB' = \frac{XY^2 + B'Y^2 - B'X^2}{2(XY)(B'Y)} = \frac{10a}{6\sqrt{6}a} = \frac{5}{3\sqrt{6}}.$$
But $\sin \angle XYB' = \sqrt{(1 - \cos^2 \angle XYB')} = \sqrt{(1 - \frac{25}{54})}$,
$$\sin \angle XYB' = \sqrt{\tfrac{29}{54}}.$$

(c) Area of triangle $XYB' = \frac{1}{2}(B'Y)(XY)\sin \angle XYB'$
$$= \frac{1}{2}(3a)(\sqrt{6}a)\sqrt{\tfrac{29}{54}}$$
$$= \frac{\sqrt{29}}{2}a^2.$$

8.8 For each of the following expressions state whether or not it is periodic, and, if it is periodic, give the period:

(a) $\dfrac{\sin x}{x}$, $x \neq 0$; (b) $\sin^2 x$; (c) $|\cos x|$.

● (a) $\dfrac{\sin x}{x}$ is not periodic.

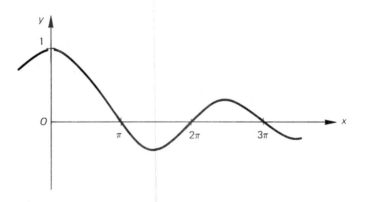

(b) $\sin^2 x$ is periodic, period π.

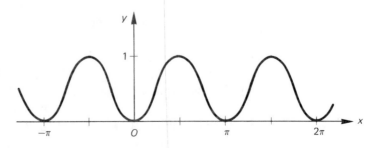

(c) $|\cos x|$ is periodic, period π.

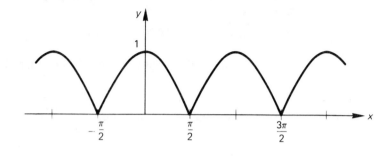

8.9 *ABCD* is one face of a cube and *AA'*, *BB'*, *CC'* and *DD'* are edges. The point *E* divides *AA'* internally in the ratio 2 : 3. Find
(a) the angle between *CE* and *C'D'*,
(b) the angle between the planes *BCE* and *ABC*.

- (a) Lines *CE* and *C'D'* are skew. The angle between them is the same as that between *CE* and any line parallel to *C'D'*, such as *CD*.

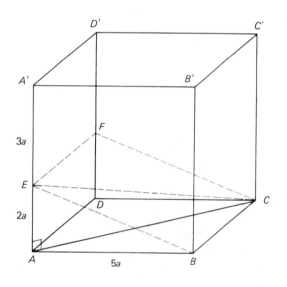

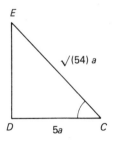

Let the length of each edge be $5a$ (since *AA'* has to be divided internally in the ratio 2 : 3).
The length of the diagonal of a face is $5\sqrt{2}a$.
In triangle *EAC*, $EA = 2a$, $AC = 5\sqrt{2}a$, $\angle EAC = 90°$.
Hence $EC^2 = (2a)^2 + (5\sqrt{2}a)^2 = 54a^2$, $EC = \sqrt{(54)}a$.
In triangle *ECD*, $\angle EDC = 90°$, $\Rightarrow$ $\cos C = \dfrac{5a}{\sqrt{(54)}a}$, $\quad \angle C = 47.1°$.
Therefore the angle between *CE* and *C'D'* is $47.1°$.

(b) The angle between planes *BCE* and *ABC* is $\angle EBA$.

$$\tan EBA = \frac{EA}{AB} = \frac{2a}{5a}, \qquad \angle EBA = 21.8°.$$

Therefore the angle between the planes *BCE* and *ABC* is $21.8°$.

8.10 A chord PQ of a circle centre O divides the circle into two segments whose areas are in the ratio $1 : 2$. If the obtuse angle POQ is α, show that
$3\alpha = 2\pi + 3\sin\alpha$.
Use a graphical method to find α.

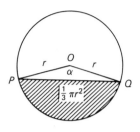

- Area of segment = area of sector OPQ − area of triangle OPQ
$$= \tfrac{1}{2}r^2\alpha - \tfrac{1}{2}r^2\sin\alpha.$$

Area of segment $= \tfrac{1}{3}$ area of circle $= \tfrac{1}{3}\pi r^2$.
Therefore $\tfrac{1}{2}r^2\alpha - \tfrac{1}{2}r^2\sin\alpha = \tfrac{1}{3}\pi r^2$.
Multiplying by $6/r^2$ gives

$$3\alpha - 3\sin\alpha = 2\pi \qquad \text{or} \qquad 3\alpha = 2\pi + 3\sin\alpha.$$

From the graphs of $y = 3\sin\alpha$ and $y = 3\alpha - 2\pi$,
$\alpha = 2.61$ radians.

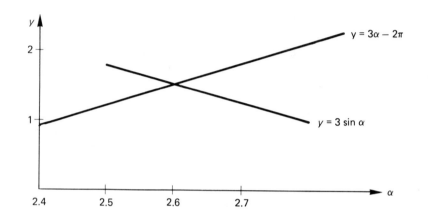

8.3 Exercises

8.1 The dimensions of triangle ABC are $AB = 7$ cm, $BC = 4$ cm and $\angle ACB = 120°$.
Length b is given by:
A, $b^2 = 65 - 56\cos 120°$;   B, $b^2 + 4b + 33 = 0$;   C, $b^2 - 4b + 33 = 0$;
D, $b^2 = 65 + 56\cos 60°$;   E, $b^2 + 4b - 33 = 0$.

8.2 O is the centre of a circle and OP and OQ are radii. If $OP = 3$ and $PQ = 5$ then the value of $\sin \angle POQ$ is

A, $\dfrac{3}{5}$;   B, $\dfrac{4}{5}$;   C, $\dfrac{7}{18}$;   D, $\dfrac{5\sqrt{11}}{18}$;   E, none of these.

8.3 The period of $\tan(3\theta + 30°)$, $\theta \in R$, is:
A, $120°$;   B, $180°$;   C, $60°$;   D, $90°$;   E, $540°$.

8.4 The four graphs shown are all drawn to the same scale. Graph (i) has equation $y = \sin x°$. State the values of the numbers a and b.
Give the equations of the graphs (ii), (iii) and (iv).

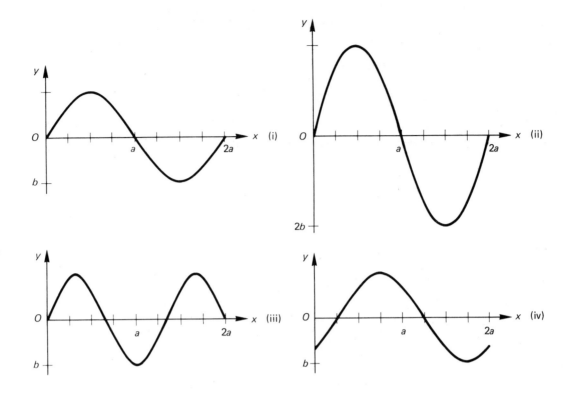

8.5 $ABCD$ is the square base of side $2a$, of a pyramid with vertex V. If $VA = VB = VC = VD = 3a$ find

(a) the vertical height of the pyramid,

(b) the angle between VA and the horizontal plane,

(c) the angle plane VAB makes with the horizontal plane,

(d) the angle between planes VAB and VAC,

(e) the angle between VA and BC.

8.6 With the usual notation for triangle ABC prove that

$$\cos A = \frac{b^2 + c^2 - a^2}{2bc}$$

(assume that the triangle is acute-angled).

If two circles of radii 5 cm and 7 cm, centres C_1 and C_2 respectively, intersect at A and B, and $C_1 C_2 = 10$ cm, calculate

(a) $\angle AC_1 C_2$;

(b) area of triangle $AC_1 C_2$;

(c) area common to both circles.

Give all answers to one decimal place.

8.7 In triangle ABC, $AB = 5$ cm, $BC = 6$ cm, $\angle ACB = 41°$. Show that there are two possible triangles, $A_1 BC$ and $A_2 BC$. Find the length of $A_1 A_2$.

8.8 $ABCD$ is a rectangle with $AB = 7$ cm, $BC = 5$ cm. AA', BB', CC' and DD' are perpendicular to $ABCD$, $AA' = BB' = 4$ cm, $CC' = DD' = 6$ cm. $A'B'C'D'$ are all on the same side of $ABCD$. If M is the midpoint of AB and N is the midpoint of $B'C'$, find $\angle D'MN$ and the area of triangle $D'MN$.

87

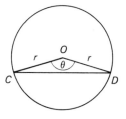

8.9 In the diagram, the chord CD cuts the circle, radius r, centre O, into two segments. CD subtends an angle θ at the centre of the circle where $0 < \theta < \pi$. Write down an expression for the perimeter P of the minor segment. If the perimeter P is half of the circumference of the circle, show that $2 \sin \dfrac{\theta}{2} = (\pi - \theta)$.

Use a graphical method to find the value of θ.

8.4 Outline Solutions to Exercises

8.1 Use the cosine rule: $49 = b^2 + 16 - 8b \cos 120°$.
$b^2 + 4b - 33 = 0$.

Answer E

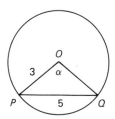

8.2 $OQ = 3$, $\cos \alpha = \dfrac{9 + 9 - 25}{18} = -\dfrac{7}{18}$;

$\sin^2 \alpha = 1 - \cos^2 \alpha = 1 - \dfrac{49}{324} = \dfrac{275}{324}$;

$\Rightarrow \quad \sin \alpha = \dfrac{5\sqrt{11}}{18}$.

Answer D

8.3 Period of $\tan(n\theta) = \dfrac{1}{n}$ (period of $\tan \theta$)

Period of $\tan(3\theta + 30°) = \dfrac{1}{3}(180°) = 60°$.

Answer C

8.4 $a = 180$, $b = -1$.
(ii) $2 \sin x°$; (iii) $\sin \frac{3}{2}x$ (notice $1\frac{1}{2}$ cycles in $360°$);
(iv) $\sin(x - 45°)$.

8.5

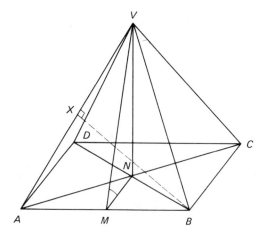

(a) $AC = BD = 2\sqrt{2}a$, height $VN = \sqrt{(9a^2 - 2a^2)} = a\sqrt{7}$.

(b) Angle VAN: $\cos \angle VAN = AN/VA = \dfrac{\sqrt{2}}{3}$, $\angle VAN = 61.9°$.

(c) M is the midpoint of AB. Angle between VAB and the horizontal is $\angle VMN$.
    $\tan \angle VMN = VN/NM = \sqrt{7}$; $\angle VMN = 69.3°$.
(d) Angle between VAB and VAC is $\frac{1}{2}$ (angle between VAB and VAD)
    $= \frac{1}{2}(\angle BXD)$ where X is the foot of the perpendicular from B to VA.

$$BX = 2a \sin \angle VAB = 2a \, \frac{\sqrt{8}}{3} = \frac{4\sqrt{2}}{3} \, a.$$

From triangle BXN, $\sin \angle BXN = BN/BX = \frac{3}{4}$. $\angle BXN = 48.6°$.

(e) Angle between VA and BC = angle between VA and AD.
    Angle $VAD = 70.5°$.

8.6 By the theorem of Pythagoras:
Triangle BAD: $BD^2 = c^2 - c^2 \cos^2 A$.
Triangle BCD: $BD^2 = a^2 - (b - c \cos A)^2$.
Equating gives $a^2 = b^2 + c^2 - 2bc \cos A$.

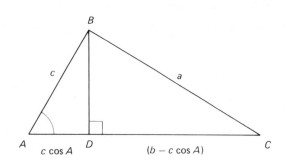

(a) Cosine rule: $\cos \angle AC_1 C_2 = \dfrac{25 + 100 - 49}{100}$, $\angle AC_1 C_2 = 40.5°$.

(b) Area of triangle $AC_1 C_2 = \frac{1}{2}(5)(10) \sin (40.5)° = 16.2 \text{ cm}^2$.

(c) Sine rule: $\dfrac{\sin \angle AC_2 C_1}{5} = \dfrac{\sin \angle AC_1 C_2}{7}$, $\angle AC_2 C_1 = 27.7°$.

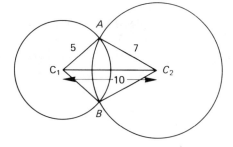

Circle centre C_1: area sector $AC_1 B = \pi(5)^2 \, \dfrac{\angle AC_1 B}{360} = 17.67 \text{ cm}^2$.

Circle centre C_2: area sector $AC_2 B = \pi(7)^2 \, \dfrac{\angle AC_2 B}{360} = 23.69 \text{ cm}^2$.

Area common to both circles = sum of sectors − area $AC_1 BC_2 = 8.9 \text{ cm}^2$.

8.7 By cosine rule, $BA^2 = BC^2 + AC^2 - 2(BC)(AC)\cos C$.
$\Rightarrow \quad AC^2 - 9.057(AC) + 11 = 0$, $AC = 7.61$ or 1.45, $A_1C = 1.45$, $A_2C = 7.61$, $A_1A_2 = 6.16$ cm.

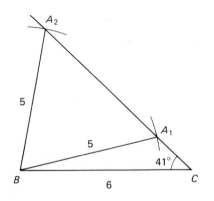

8.8 From right-angled triangles, $B'C' = \sqrt{29}$, $C'N = \frac{1}{2}\sqrt{29}$, $D'N = 7.5$.
$PN = 5$, $PM = \frac{1}{2}\sqrt{74} = 4.301$, $MN = 6.595$, $MD = 6.103$, $MD' = 8.559$.

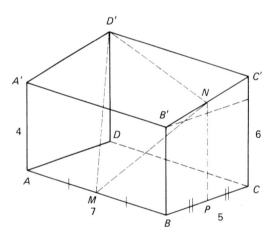

By cos rule, triangle MND', $\cos \angle D'MN = \dfrac{NM^2 + MD'^2 - D'N^2}{2(NM)(MD')}$, $\angle D'MN = 57.6°$.

Area of triangle $D'MN = \frac{1}{2}(MD')(MN)\sin \angle NMD' = 23.8$ cm^2.

8.9 From the diagram (see Q. 8.9) chord $CD = (2r) \sin \dfrac{\theta}{2}$, minor arc $CD = r\theta$.

Perimeter of minor segment $= (2r) \sin \dfrac{\theta}{2} + r\theta = \pi r$.

$$\Rightarrow \quad 2 \sin \frac{\theta}{2} = \pi - \theta.$$

From the graphs of $y = 2 \sin \dfrac{\theta}{2}$ and $y = \pi - \theta$, $\theta = 1.66$ radians.

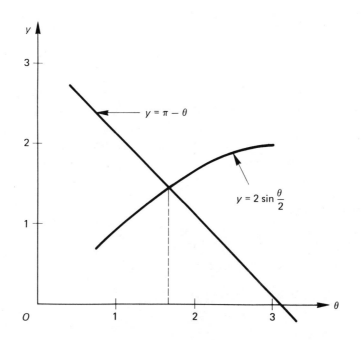

9 Trigonometric Identities and Equations

Knowledge and use of the formulae for sin $(A \pm B)$, cos $(A \pm B)$, tan $(A \pm B)$, sin $A \pm$ sin B etc. Identities such as $\sin^2 A + \cos^2 A \equiv 1$. Expression of $a \cos \theta + b \sin \theta$ in the form $r \cos (\theta \pm \alpha)$.

General solution of simple trigonometric equations, including graphical interpretation.

9.1 Fact Sheet

(a) Pythagorean Identities

- $\sin^2 A + \cos^2 A \equiv 1$.
- $\tan^2 A + 1 \equiv \sec^2 A$.
- $\cot^2 A + 1 \equiv \csc^2 A$.

(b) Compound Angle Formulae

- Addition

$\sin(A \pm B) \equiv \sin A \cos B \pm \cos A \sin B$,

$\cos(A \pm B) \equiv \cos A \cos B \mp \sin A \sin B$,

$\tan(A \pm B) \equiv \dfrac{\tan A \pm \tan B}{1 \mp \tan A \tan B}$.

- Factor

$\sin A + \sin B \equiv 2 \sin\left(\dfrac{A+B}{2}\right) \cos\left(\dfrac{A-B}{2}\right)$,

$\sin A - \sin B \equiv 2 \cos\left(\dfrac{A+B}{2}\right) \sin\left(\dfrac{A-B}{2}\right)$,

$\cos A + \cos B \equiv 2 \cos\left(\dfrac{A+B}{2}\right) \cos\left(\dfrac{A-B}{2}\right)$,

$\cos A - \cos B \equiv -2 \sin\left(\dfrac{A+B}{2}\right) \sin\left(\dfrac{A-B}{2}\right)$.

- Product

$\sin A \cos B \equiv [\sin(A+B) + \sin(A-B)]/2$,

$\cos A \sin B \equiv [\sin(A+B) - \sin(A-B)]/2$,

$\cos A \cos B \equiv [\cos(A+B) + \cos(A-B)]/2$,

$\sin A \sin B \equiv [\cos(A-B) - \cos(A+B)]/2$,

(c) Multiple Angle

- $\sin 2A \equiv 2 \sin A \cos A$.
- $\cos 2A \equiv \cos^2 A - \sin^2 A$
  $\equiv 2 \cos^2 A - 1 \equiv 1 - 2 \sin^2 A$.
- $\tan 2A \equiv \dfrac{2 \tan A}{1 - \tan^2 A}$.
- $\cos^2 A \equiv \dfrac{1 + \cos 2A}{2}$.
- $\sin^2 A \equiv \dfrac{1 - \cos 2A}{2}$.
- $\sin 3A \equiv 3 \sin A - 4 \sin^3 A$.
- $\cos 3A \equiv 4 \cos^3 A - 3 \cos A$.

(d) (r, α) Formula

- If a and b are positive, $a \cos \theta \pm b \sin \theta \equiv r \cos (\theta \mp \alpha_1)$
  $b \sin \theta \pm a \cos \theta \equiv r \sin (\theta \pm \alpha_2)$.
- $r^2 = a^2 + b^2$, $\tan \alpha_1 = \dfrac{b}{a}$, $\tan \alpha_2 = \dfrac{a}{b}$ (r positive, α_1 and α_2 acute).

(e) 't' Formulae

- If $t = \tan \dfrac{x}{2}$, then $\sin x \equiv \dfrac{2t}{1 + t^2}$, $\cos x \equiv \dfrac{1 - t^2}{1 + t^2}$, $\tan x \equiv \dfrac{2t}{1 - t^2}$.

(f) Hints for Solutions

(i) When in doubt, or if more than two trigonometric functions are present, change all of the functions into sines and cosines.

(ii) Compare left-hand side and right-hand side. If the angles are all the same, e.g. all θ, use the basic identities from Chapter 8 and the Pythagorean identities. If the angles are different, use the compound angle formulae.

(iii) In identity questions, start with the more complicated side, and work on one side at a time.

9.2 Worked Examples

9.1 If $2 \sin \theta - 3 \cos \theta$ is expressed in the form $r \cos (\theta - \alpha)$, where $r > 0$ and $0 \leqslant \alpha \leqslant 2\pi$, then α lies between

A, 0 and $\pi/2$;   B, $\pi/2$ and π;   C, π and $3\pi/2$;   D, $3\pi/2$ and 2π.

- $$2 \sin \theta - 3 \cos \theta \equiv r \cos (\theta - \alpha)$$
  $$\equiv r \cos \theta \cos \alpha + r \sin \theta \sin \alpha.$$

93

Equating coefficients of $\sin \theta$, $2 \equiv r \sin \alpha$.
Equating coefficients of $\cos \theta$, $-3 \equiv r \cos \alpha$.
Since $r > 0$, $\sin \alpha > 0$ and $\cos \alpha < 0$, α must lie in the second quadrant.

Answer B

9.2 Given that $5 \cos \theta - 12 \sin \theta = 13 \cos (\theta + 67.4°)$, which of the following equations has/have solutions for $\theta \in \mathbb{R}$?
(a) $5 \cos \theta - 12 \sin \theta = 6$; (b) $5 \cos \theta - 12 \sin \theta = -10$;
(c) $5 \cos \theta - 12 \sin \theta = 17$; (d) $5 \cos \theta - 12 \sin \theta = -13$.

A, (a) only; B, (a) and (b) only; C, (c) only;
D, (a), (b) and (d) only; E, (b), (c) and (d) only.

- Since $f(\theta) = 5 \cos \theta - 12 \sin \theta = 13 \cos (\theta + 67.4)$, then $f(\theta)$ has a maximum value of 13 and a minimum value of -13 $\Rightarrow$ $-13 \leqslant f(\theta) \leqslant 13$.
Equations a, b and d satisfy this condition. **Answer D**

9.3
(a) Prove that $\cos 3\theta + \sin 3\theta = (\cos \theta - \sin \theta)(1 + 2 \sin 2\theta)$.
(b) If $2a \cos 2x + b \sin 2x + 2c = 0$, where $a \neq 0$ and $a \neq c$, find an equation for $\tan x$.

State the sum and product of the roots of this equation, $\tan x_1$ and $\tan x_2$, and hence deduce that $\tan (x_1 + x_2) = \dfrac{b}{2a}$.

- (a) To prove
$$\cos 3\theta + \sin 3\theta = (\cos \theta - \sin \theta)(1 + 2 \sin 2\theta):$$
$$\cos 3\theta = 4 \cos^3 \theta - 3 \cos \theta, \quad \sin 3\theta = 3 \sin \theta - 4 \sin^3 \theta.$$
l.h.s.: $\cos 3\theta + \sin 3\theta$
$$= 4 \cos^3 \theta - 3 \cos \theta + 3 \sin \theta - 4 \sin^3 \theta$$
$$= 4(\cos^3 \theta - \sin^3 \theta) - 3(\cos \theta - \sin \theta)$$
$$= 4(\cos \theta - \sin \theta)(\cos^2 \theta + \cos \theta \sin \theta + \sin^2 \theta) - 3(\cos \theta - \sin \theta)$$
$$= (\cos \theta - \sin \theta)(4 + 4 \cos \theta \sin \theta - 3),$$
$$\Rightarrow \quad \cos 3\theta + \sin 3\theta = (\cos \theta - \sin \theta)(1 + 2 \sin 2\theta).$$

(b) Given: $2a \cos 2x + b \sin 2x + 2c = 0.$
$$\cos 2x = \frac{1 - t^2}{1 + t^2} \quad \text{and} \quad \sin 2x = \frac{2t}{1 + t^2}, \text{ where } t = \tan x.$$

Substituting: $2a \dfrac{(1 - t^2)}{1 + t^2} + \dfrac{2bt}{1 + t^2} + 2c = 0.$

Multiply by $(1 + t^2)$: $2a - 2at^2 + 2bt + 2c + 2ct^2 = 0$
$$t^2 (2c - 2a) + 2bt + (2a + 2c) = 0$$
$$t^2 (c - a) + bt + (a + c) = 0.$$

Sum of roots: $\tan x_1 + \tan x_2 = \dfrac{-b}{(c - a)}.$

Product of roots: $\tan x_1 \tan x_2 = \dfrac{(a + c)}{(c - a)}.$

$$\tan(x_1 + x_2) = \frac{\tan x_1 + \tan x_2}{1 - \tan x_1 \tan x_2} = \frac{-\dfrac{b}{(c-a)}}{1 - \dfrac{a+c}{c-a}}$$

$$= \frac{-b}{(c-a)-(a+c)} = \frac{b}{2a}.$$

9.4 Given that $\sin(A - B) \neq 0$, prove that

$$\frac{\sin A + \sin B}{\sin(A - B)} = \sin\left(\frac{A + B}{2}\right) \operatorname{cosec}\left(\frac{A - B}{2}\right).$$

Use this result to find cosec $15°$ in surd form.

- To prove $\dfrac{\sin A + \sin B}{\sin(A - B)} = \sin\left(\dfrac{A + B}{2}\right) \operatorname{cosec}\left(\dfrac{A - B}{2}\right)$:

  The product on the r.h.s. suggests the use of the factor formula for the numerator on the l.h.s.

  $\sin A + \sin B = 2 \sin\left(\dfrac{A + B}{2}\right) \cos\left(\dfrac{A - B}{2}\right)$ (factor formula);

  $\sin(A - B) = 2 \sin\left(\dfrac{A - B}{2}\right) \cos\left(\dfrac{A - B}{2}\right)$ (double angle formula).

  Therefore $\dfrac{\sin A + \sin B}{\sin(A - B)} = \dfrac{2 \sin\left(\dfrac{A + B}{2}\right) \cos\left(\dfrac{A - B}{2}\right)}{2 \sin\left(\dfrac{A - B}{2}\right) \cos\left(\dfrac{A - B}{2}\right)}$

  $$= \frac{\sin\left(\dfrac{A + B}{2}\right)}{\sin\left(\dfrac{A - B}{2}\right)}$$

  $$= \sin\left(\frac{A + B}{2}\right) \operatorname{cosec}\left(\frac{A - B}{2}\right).$$

If $\dfrac{A - B}{2} = 15°$ then $A - B = 30°$.

Choose values for A and B which satisfy this condition.
Let $A = 60°$, $B = 30°$, $A + B = 90°$.

Then $\dfrac{\sin 60° + \sin 30°}{\sin 30°} = \sin 45° \operatorname{cosec} 15°$

$$\frac{\dfrac{\sqrt{3}}{2} + \dfrac{1}{2}}{\dfrac{1}{2}} = \frac{1}{\sqrt{2}} \operatorname{cosec} 15°$$

$$\sqrt{2}\,(\sqrt{3} + 1) = \operatorname{cosec} 15°$$

$$\Rightarrow \quad \operatorname{cosec} 15° = \sqrt{6} + \sqrt{2}.$$

95

9.5 Find, without using tables or a calculator, the value of

$$\left(\sin \frac{5\pi}{12} - \cos \frac{5\pi}{12} \right)^2.$$

● $$\left(\sin \frac{5\pi}{12} - \cos \frac{5\pi}{12} \right)^2 = \sin^2 \frac{5\pi}{12} - 2 \sin \frac{5\pi}{12} \cos \frac{5\pi}{12} + \cos^2 \frac{5\pi}{12}$$

$$= \left(\sin^2 \frac{5\pi}{12} + \cos^2 \frac{5\pi}{12} \right) - \sin \frac{5\pi}{6}$$

$$= 1 - \tfrac{1}{2}$$

$$= \tfrac{1}{2}.$$

9.6 Find expressions for $\dfrac{\sin 2\theta}{\sin \theta}$, $\dfrac{\sin 3\theta}{\sin \theta}$ and $\dfrac{\sin 4\theta}{\sin \theta}$ in terms of $\cos \theta$.

● $\sin 2\theta = 2 \sin \theta \cos \theta.$ (1)

$\sin 3\theta = 3 \sin \theta - 4 \sin^3 \theta = \sin \theta (3 - 4 \sin^2 \theta)$

$\quad\quad\quad = \sin \theta \{3 - 4(1 - \cos^2\theta)\}$

$\quad\quad\quad = \sin \theta (4 \cos^2 \theta - 1).$ (2)

$\sin 4\theta = 2 \sin 2\theta \cos 2\theta = 4 \sin \theta \cos \theta (2 \cos^2 \theta - 1).$ (3)

From 1, $\dfrac{\sin 2\theta}{\sin \theta} = 2 \cos \theta.$

From 2, $\dfrac{\sin 3\theta}{\sin \theta} = 4 \cos^2 \theta - 1.$

From 3, $\dfrac{\sin 4\theta}{\sin \theta} = 4 \cos \theta (2 \cos^2 \theta - 1) = 8 \cos^3 \theta - 4 \cos \theta.$

9.7 Find the angles between $-180°$ and $180°$ which satisfy the equations

(a) $3 \sin^2 \theta - 2 \cos^2 \theta = 1,$

(b) $3 \sin \theta - 2 \cos \theta = 1.$

● (a) $3 \sin^2 \theta - 2 \cos^2 \theta = 1.$

Using the identity $\sin^2 \theta + \cos^2 \theta = 1,$

$$\cos^2 \theta = 1 - \sin^2 \theta.$$

Substituting: $3 \sin^2 \theta - 2 + 2 \sin^2 \theta = 1,$

$$5 \sin^2 \theta = 3 \quad \Rightarrow \quad \sin \theta = \pm\sqrt{\tfrac{3}{5}}.$$

When $\sin \theta = \sqrt{\tfrac{3}{5}}$, $\theta = (50.8° \text{ or } 180° - 50.8°) + 360n°.$

When $\sin \theta = -\sqrt{\tfrac{3}{5}}$, $\theta = (-50.8° \text{ or } -180° + 50.8°) + 360n°.$

Therefore $\theta = -129.2°, -50.8°, 50.8°, 129.2°.$

(b) **There are two standard methods for this question,**
(i) 't' substitution, (ii) (R, α).

(i) 't' substitution: $\sin \theta = \dfrac{2t}{1 + t^2}$, $\cos \theta = \dfrac{1 - t^2}{1 + t^2}$, where $t = \tan \dfrac{\theta}{2}.$

Substituting: $\dfrac{6t}{1 + t^2} - \dfrac{2(1 - t^2)}{1 + t^2} = 1$

$$6t - 2 + 2t^2 = 1 + t^2$$

$$t^2 + 6t - 3 = 0.$$

Solving: $t = 0.4641$ or -6.4641.

When $\tan \dfrac{\theta}{2} = 0.4641$, $\dfrac{\theta}{2} = 24.9° + 180n°$, $\theta = 49.8° + 360n°$.

When $\tan \dfrac{\theta}{2} = -6.4641$, $\dfrac{\theta}{2} = -81.2° + 180n°$, $\theta = -162.4° + 360n°$.

Therefore $\theta = -162.4°$ and $49.8°$ in the range $-180°$ to $180°$.

(ii) Let $3 \sin \theta - 2 \cos \theta \equiv R \sin (\theta - \alpha)$
$$\equiv R \sin \theta \cos \alpha - R \cos \theta \sin \alpha.$$
Comparing terms in $\sin \theta$, $3 = R \cos \alpha$.
Comparing terms in $\cos \theta$, $2 = R \sin \alpha$.
By division, $\tan \alpha = \frac{2}{3}$, $\therefore \alpha = 33.7°$.
$R^2 = 3^2 + 2^2 \Rightarrow R = \sqrt{13}$;
Therefore $\sqrt{13} \sin (\theta - 33.7°) = 1$
$$\sin (\theta - 33.7°) = \frac{1}{\sqrt{13}} = 0.2774.$$
$\theta - 33.7° = (16.1°$ or $163.9°) + 360n°$.
Therefore $\theta = (49.8°$ or $197.6°) + 360n°$.
Therefore $\theta = -162.4°$ and $49.8°$ in the range $-180°$ to $180°$.

9.8 Solve the equations
(a) $3 \sin x - 2 \sin 4x + 3 \sin 7x = 0$ for $0° \leqslant x \leqslant 180°$.
(b) $3 \sin 2x = 2 \tan x$ for $-180° \leqslant x \leqslant 180°$.
(c) $8 \sin^3 x - 6 \sin x = 1$ for $0° < x < 360°$.

● (a) $3 \sin x - 2 \sin 4x + 3 \sin 7x = 0$.

Since the first and third terms have the same coefficients, combine them and use the factor formula:

$3 (\sin x + \sin 7x) = 6 \sin 4x \cos 3x$;
substituting: $6 \sin 4x \cos 3x - 2 \sin 4x = 0$
$$2 \sin 4x (3 \cos 3x - 1) = 0$$
$\Rightarrow \sin 4x = 0$ or $3 \cos 3x = 1$, $\cos 3x = \frac{1}{3}$.
When $\sin 4x = 0$, $4x = 180n°$, $x = 45n°$ (n is an integer).
When $\cos 3x = \frac{1}{3}$, $3x = 360n° \pm 70.5°$, $x = 120n° \pm 23.5°$.
For $0° \leqslant x \leqslant 180°$, $x = 0°$, $23.5°$, $45°$, $90°$, $96.5°$, $135°$, $143.5°$, $180°$.

(b) $3 \sin 2x = 2 \tan x$.
Use 't' substitution:
$$\sin 2x = \frac{2t}{1 + t^2} \qquad \text{where } t = \tan x.$$
Substituting: $\dfrac{6t}{1 + t^2} = 2t \Rightarrow 6t = 2t + 2t^3 \Rightarrow 2t^3 - 4t = 0$
$$2t (t^2 - 2) = 0 \Rightarrow t = 0, \sqrt{2} \text{ or } -\sqrt{2}.$$
When $\tan x = 0$, $x = 180n°$.
When $\tan x = \sqrt{2}$, $x = 54.7° + 180n°$.
When $\tan x = -\sqrt{2}$, $x = -54.7° + 180n°$.
For $-180° \leqslant x \leqslant 180°$,
$x = -180°$, $-125.3°$, $-54.7°$, $0°$, $54.7°$, $125.3°$, $180°$.

(c) $8 \sin^3 x - 6 \sin x = 1$.

$\sin 3x = 3 \sin x - 4 \sin^3 x \quad \Rightarrow \quad 8 \sin^3 x - 6 \sin x = -2 \sin 3x$.

Substituting into the equation: $-2 \sin 3x = 1$, $\sin 3x = -\frac{1}{2}$.

When $\sin 3x = -\frac{1}{2}$, $3x = -30° + 360n°$ or $210° + 360n°$

$\Rightarrow \quad x = -10° + 120n°$ or $70° + 120n°$.

In the interval, $0° < x < 360°$, $x = 70°, 110°, 190°, 230°, 310°, 350°$.

9.9

(a) Find the general solution, in degrees, of the equation
$$\sin x + \sin 2x + \sin 3x = \sin x \sin 2x.$$

(b) Find the general solution, in radians, of the equation
$$\sin 2\theta + \cos 2\theta = \sin \theta - \cos \theta + 1.$$

● (a) By the factor formula,
$$\sin x + \sin 3x = 2 \sin 2x \cos x.$$

Equation becomes
$$2 \sin 2x \cos x + \sin 2x - \sin x \sin 2x = 0.$$
$$\Rightarrow \quad \sin 2x (2 \cos x - \sin x + 1) = 0.$$

Either $\sin 2x = 0$, or $2 \cos x - \sin x + 1 = 0$.

When $\sin 2x = 0$, $2x = 180n°$, so $x = 90n°$.

When $2 \cos x - \sin x + 1 = 0$, then use either 't-substitutions' or compound angles.

$2 \cos x - \sin x = r \cos(x + \alpha)$, where $r = \sqrt{(2^2 + 1^2)}$, and $\tan \alpha = 0.5$,
so $\alpha = 26.6°$.

$$\sqrt{5} \cos (x + 26.6°) = -1 \quad \Rightarrow \quad \cos (x + 26.6°) = -\frac{1}{\sqrt{5}}$$

$\Rightarrow \quad x + 26.6° = 360n° \pm 116.6°$,

$x = 360n° + 90°$ or $360n° - 143.1°$.

$x = 360n° + 90°$ is included in the earlier solution $x = 90n°$.

Hence general solution: $x = 90n°$ or $360n° - 143.1°$ (n is any integer).

(b) $\sin 2\theta + \cos 2\theta = \sin \theta - \cos \theta + 1$

Hint: It would be convenient to get rid of the +1 on the r.h.s.
Either change $\cos 2\theta$ to $1 - 2\sin^2 \theta$
or change $\cos 2\theta$ to $\cos^2 \theta - \sin^2 \theta$ and 1 to $\cos^2 \theta + \sin^2 \theta$.

$\Rightarrow \quad 2 \sin \theta \cos \theta + 1 - 2 \sin^2 \theta = \sin \theta - \cos \theta + 1$

(or $2 \sin \theta \cos \theta + \cos^2 \theta - \sin^2 \theta = \sin \theta - \cos \theta + \cos^2 \theta + \sin^2 \theta$),

giving $2 \sin \theta \cos \theta - 2 \sin^2 \theta + \cos \theta - \sin \theta = 0$

$\Rightarrow \quad 2 \sin \theta (\cos \theta - \sin \theta) + (\cos \theta - \sin \theta) = 0$

$\Rightarrow \quad (\cos \theta - \sin \theta)(2 \sin \theta + 1) = 0$.

Either

$$2 \sin \theta + 1 = 0 \quad \Rightarrow \quad \sin \theta = -0.5 \quad \Rightarrow \quad \theta = n\pi + (-1)^n \left(-\frac{\pi}{6}\right),$$

or $\cos \theta - \sin \theta = 0 \quad \Rightarrow \quad \tan \theta = 1 \quad \Rightarrow \quad \theta = n\pi + \frac{\pi}{4}$.

General solution: $\theta = n\pi + (-1)^{n+1} \left(\frac{\pi}{6}\right)$ or $(4n + 1) \frac{\pi}{4}$.

9.10 Obtain the general solution of the equation
$$\sin \theta° = \cos \alpha°$$
for θ in terms of α.

- $\cos \alpha° = \sin(90° - \alpha°)$ or $\sin(90° + \alpha°)$.
  Therefore, $\sin \theta° = \sin(90° - \alpha°)$ or $\sin(90° + \alpha°)$.
  $\Rightarrow$ $\theta = 360n + (90 - \alpha)$ or $360n + (90 + \alpha)$ where n is any integer.
  $\Rightarrow$ $\theta = (4n + 1)90 \pm \alpha$.

9.3 Exercises

9.1 The solution of the equation $3 \sin x = -\sqrt{3} \cos x$, where $-\pi \leqslant x \leqslant \dfrac{\pi}{2}$, is

A, $-\pi$;   B, $-\dfrac{\pi}{2}$;   C, $-\dfrac{\pi}{6}$;   D, $\dfrac{\pi}{6}$;   E, $\dfrac{\pi}{3}$.

9.2 Show that $\operatorname{cosec} x + \cot x = \cot \dfrac{x}{2}$.

Deduce the exact values of $\tan 15°$ and $\tan 67\frac{1}{2}°$ in rational surd form.

9.3 Given that $0° \leqslant \theta \leqslant 360°$, express $4 \sin \theta - 3 \cos \theta$ in the form $r \sin(\theta + \alpha)$, where r is positive and $-180° \leqslant \alpha \leqslant 180°$. Hence write down the greatest and least values of this expression and state the corresponding values of θ to the nearest $0.1°$.

9.4

(a) Prove that (i) $\sin 3x = 4 \sin x \sin \left(\dfrac{\pi}{3} + x \right) \sin \left(\dfrac{\pi}{3} - x \right)$,

(ii) $\tan 3x = \tan x \tan \left(\dfrac{\pi}{3} + x \right) \tan \left(\dfrac{\pi}{3} - x \right)$.

(b) Prove the formulae $\sin \theta = \dfrac{2t}{1 + t^2}$ and $\cos \theta = \dfrac{1 - t^2}{1 + t^2}$ where $t = \tan \dfrac{\theta}{2}$.

Hence or otherwise prove that $\tan A = \operatorname{cosec} 2A - \cot 2A$.

9.5 Given that $3 \cos x + 2 \sec x + 5 = 0$, find, without using tables or calculator, all the possible values of $\sin x$ and of $\tan^2 x$.

9.6 Prove that $\left(\dfrac{\sin 5A}{\sin A} \right)^2 - \left(\dfrac{\cos 5A}{\cos A} \right)^2 = 8 \cos 2A (4 \cos^2 2A - 1)$.

9.7 Prove that $\sin 3A = 3 \sin A - 4 \sin^3 A$. Given that $\sin 3x = \sin^2 x$, find three possible values for $\sin x$.

Hence find all of the solutions of the equation $\sin 3x = \sin^2 x$ for $90° \leqslant x \leqslant 270°$. Using the same axes, sketch the graphs of the functions $\sin 3x$ and $\sin^2 x$ for $90° \leqslant x \leqslant 270°$. Find the subset of values of x for which $\sin 3x < \sin^2 x$.

9.8 Given that $a \cos x - b \sin x \equiv 2 \cos \left(\dfrac{\pi}{6} + x \right)$, find the constants a and b.

Write down the general solution of the equation $\sqrt{6} \cos x - \sqrt{2} \sin x = 2$, giving your answer in radians. (L)

9.9 Solve the equations
(a) $\cos 3x + 2 \cos x = 0$,   for $0° \leqslant x \leqslant 360°$;
(b) $4 \cos x = \operatorname{cosec} x$,   for $-\pi \leqslant x \leqslant \pi$.

9.10 Find the general solutions, in degrees, of
(a) $\sin 3x - \sin x = \frac{1}{2} \cos 2x$,
(b) $4 \cot 2x = \tan x$.

9.4   Outline Solutions to Exercises

9.1 $3 \sin x = -\sqrt{3} \cos x$; dividing through by $3 \cos x$ gives

$$\tan x = \frac{-1}{\sqrt{3}} \quad \Rightarrow \quad x = -\frac{\pi}{6}.$$

Answer C

9.2   Using $t = \tan \dfrac{x}{2}$ gives $\operatorname{cosec} x = \dfrac{1 + t^2}{2t}$, $\cot x = \dfrac{1 - t^2}{2t}$,

so $\operatorname{cosec} x + \cot x = \dfrac{1 + t^2}{2t} + \dfrac{1 - t^2}{2t} = \dfrac{1}{t} = \cot \dfrac{x}{2}$.

Put $x = 30°$: $\operatorname{cosec} 30° + \cot 30° = \cot 15°$. But $\operatorname{cosec} 30° = 2$,

$$\cot 30° = \sqrt{3} \quad \Rightarrow \quad \tan 15° = \frac{1}{2 + \sqrt{3}} = \frac{2 - \sqrt{3}}{(2 + \sqrt{3})(2 - \sqrt{3})} = 2 - \sqrt{3}.$$

Put $x = 135°$: $\operatorname{cosec} 135° + \cot 135° = \cot 67.5°$.

But $\operatorname{cosec} 135° = \sqrt{2}$, $\cot 135° = -1 \quad \Rightarrow \quad \tan 67.5° = \dfrac{1}{\sqrt{2} - 1} = \sqrt{2} + 1$.

9.3 $4 \sin \theta - 3 \cos \theta = r \sin(\theta + \alpha) = r \sin \theta \cos \alpha + r \cos \theta \sin \alpha$.
$\sin \theta$ terms give $4 = r \cos \alpha$,   (1)
$\cos \theta$ terms give $-3 = r \sin \alpha$.   (2)
     From 1, $\cos \alpha$ is positive, and, from 2, $\sin \alpha$ is negative, so α lies in the fourth quadrant.
$\tan \alpha = -0.75$, $\alpha = -36.9°$.
$r^2 = 4^2 + (-3)^2 \quad \Rightarrow \quad r = 5$
$\Rightarrow \quad 4 \sin \theta - 3 \cos \theta = 5 \sin(\theta - 36.9°)$.
Maximum value $= 5$,
when $\sin(\theta - 36.9°) = 1 \quad \Rightarrow \quad \theta - 36.9° = 90° \quad \Rightarrow \quad \theta = 126.9°$.
Minimum value $= -5$ when $\sin(\theta - 36.9)° = -1 \quad \Rightarrow \quad \theta = 306.9°$.

9.4
(a)   (i) $4 \sin x \sin \left(\dfrac{\pi}{3} + x \right) \sin \left(\dfrac{\pi}{3} - x \right) = 2 \sin x \left(\cos 2x - \cos \dfrac{2\pi}{3} \right)$

$$= 2 \sin x (1 - 2 \sin^2 x + 0.5)$$
$$= 3 \sin x - 4 \sin^3 x = \sin 3x.$$

(ii) Similarly, $4 \cos x \cos \left(\dfrac{\pi}{3} + x \right) \cos \left(\dfrac{\pi}{3} - x \right) = \cos 3x$.

Hence $$\frac{4 \sin x \sin\left(\frac{\pi}{3} + x\right) \sin\left(\frac{\pi}{3} - x\right)}{4 \cos x \cos\left(\frac{\pi}{3} + x\right) \cos\left(\frac{\pi}{3} - x\right)} = \frac{\sin 3x}{\cos 3x}.$$

Hence result.

(b) $\sin \theta = 2 \sin \frac{\theta}{2} \cos \frac{\theta}{2},\qquad \cos \theta = \cos^2 \frac{\theta}{2} - \sin^2 \frac{\theta}{2};$

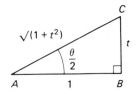

From sketch, $\tan \frac{\theta}{2} = t \;\Rightarrow\; AB = 1, BC = t \;\Rightarrow\; AC = \sqrt{(1 + t^2)}.$

$$\sin \frac{\theta}{2} = \frac{t}{\sqrt{(1 + t^2)}},\; \cos \frac{\theta}{2} = \frac{1}{\sqrt{(1 + t^2)}},$$

hence results.

$$\operatorname{cosec} 2A - \cot 2A = \frac{1}{\sin 2A} - \frac{1}{\tan 2A}$$

$$= \frac{1 + t^2}{2t} - \frac{1 - t^2}{2t} \quad \text{where } t = \tan A$$

$$= t$$

$$\Rightarrow \quad \operatorname{cosec} 2A - \cot 2A = \tan A.$$

9.5 $\quad 3 \cos x + 2 \sec x + 5 = 0 \;\Rightarrow\; 3 \cos^2 x + 2 + 5 \cos x = 0$

$$\Rightarrow \quad (3 \cos x + 2)(\cos x + 1) = 0$$

$$\Rightarrow \quad \cos x = -\tfrac{2}{3} \text{ or } -1.$$

$$\sin^2 x = 1 - \cos^2 x \;\Rightarrow\; \sin x = \pm \frac{\sqrt{5}}{3} \text{ or } 0.$$

$$\tan^2 x = \tfrac{5}{4} \text{ or } 0.$$

9.6 $\quad \left(\frac{\sin 5A}{\sin A}\right)^2 - \left(\frac{\cos 5A}{\cos A}\right)^2 = \left(\frac{\sin 5A}{\sin A} - \frac{\cos 5A}{\cos A}\right)\left(\frac{\sin 5A}{\sin A} + \frac{\cos 5A}{\cos A}\right)$

$$= \frac{\sin 5A \cos A - \cos 5A \sin A}{\sin A \cos A} \cdot \frac{\sin 5A \cos A + \cos 5A \sin A}{\sin A \cos A}$$

$$= \frac{4 \sin 4A \sin 6A}{\sin 2A \sin 2A} = \frac{8 \sin 2A \cos 2A (3 \sin 2A - 4 \sin^3 2A)}{\sin^2 2A}$$

$$= 8 \cos 2A (3 - 4 \sin^2 2A) = 8 \cos 2A (4 \cos^2 2A - 1).$$

9.7 $\quad \sin 3A = \sin 2A \cos A + \cos 2A \sin A$

$$= 2 \sin A \cos^2 A + \sin A - 2 \sin^3 A$$

$$= 3 \sin A - 4 \sin^3 A.$$

$\sin 3x = \sin^2 x \;\Rightarrow\; 3 \sin x - 4 \sin^3 x = \sin^2 x$

$$\Rightarrow \quad \sin x (4 \sin^2 x + \sin x - 3) = 0.$$

$\sin x = 0$ or 0.75 or -1.

$\sin x = 0 \;\Rightarrow\; x = 180n°.$

$\sin x = 0.75 \;\Rightarrow\; x = 48.6° + 360n°$ or $131.4° + 360n°.$

$\sin x = -1 \;\Rightarrow\; x = 270° + 360n°.$

In the range $90° \leqslant x \leqslant 270°$, $x = 131.4°, 180°, 270°.$

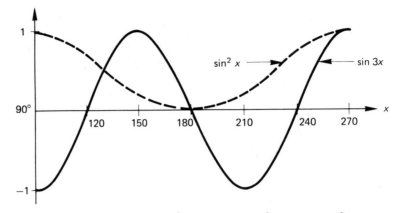

$$\sin 3x < \sin^2 x \quad \text{when} \quad 90° < x < 131.4° \quad \text{or} \quad 180° < x < 270°.$$

9.8 $a \cos x - b \sin x \equiv 2 \cos \left(\dfrac{\pi}{6} + x \right) \equiv \sqrt{3} \cos x - \sin x \quad \Rightarrow \quad a = \sqrt{3}, b = 1.$

$$\sqrt{6} \cos x - \sqrt{2} \sin x = 2 \quad \Rightarrow \quad 2\sqrt{2} \cos \left(\dfrac{\pi}{6} + x \right) = 2$$

$$\Rightarrow \quad \dfrac{\pi}{6} + x = 2n\pi \pm \dfrac{\pi}{4}$$

$$\Rightarrow \quad x = 2n\pi - \dfrac{\pi}{6} \pm \dfrac{\pi}{4}.$$

9.9

(a) $\qquad \cos 3x = 4 \cos^3 x - 3 \cos x,$

so $\cos 3x + 2 \cos x = 0$ becomes

$$\cos x (4 \cos^2 x - 1) = 0.$$
$$\cos x = 0 \quad \Rightarrow \quad x = 90° + 180n°.$$
$$\cos x = 0.5 \quad \Rightarrow \quad x = 360n° \pm 60°.$$
$$\cos x = -0.5 \quad \Rightarrow \quad x = 360n° \pm 120°.$$

In the interval $0° \leqslant x \leqslant 360°$, solutions are

$$60°, 90°, 120°, 240°, 270°, 300°.$$

(b) $\cos ec\, x = \dfrac{1}{\sin x} ,$

so $\qquad 4 \cos x = \cos ec\, x \quad \Rightarrow \quad 4 \cos x \sin x = 1$

$$\Rightarrow \quad 2 \sin 2x = 1 \quad \Rightarrow \quad 2x = n\pi + (-1)^n \dfrac{\pi}{6} \quad \Rightarrow \quad x = n\dfrac{\pi}{2} + (-1)^n \dfrac{\pi}{12}.$$

In the interval $-\pi \leqslant x \leqslant \pi$, solutions are

$$x = - \dfrac{11}{12} \pi, - \dfrac{7}{12} \pi, \dfrac{1}{12} \pi, \dfrac{5}{12} \pi.$$

9.10

(a) $\sin 3x - \sin x = 2 \cos 2x \sin x,$

so equation becomes $\cos 2x (4 \sin x - 1) = 0.$
$$\cos 2x = 0 \quad \Rightarrow \quad x = 45° + 90n°;$$
$$\sin x = 0.25 \quad \Rightarrow \quad x = (-1)^n (14.5°) + 180n°.$$

So general solution is

$$x = 45° + 90n° \quad \text{or} \quad (-1)^n (14.5°) + 180n°.$$

(b) $4 \cot 2x = \tan x \quad \Rightarrow \quad 4 = \tan x \tan 2x.$

Put $t = \tan x$, then $\tan 2x = \dfrac{2t}{1 - t^2} ,$ and equation becomes

$$4 - 4t^2 = 2t^2 \quad \Rightarrow \quad t = \pm \sqrt{(\tfrac{2}{3})}.$$

When $\tan x = \sqrt{(\tfrac{2}{3})}$, $x = 39.2° + 180n°.$
When $\tan x = -\sqrt{(\tfrac{2}{3})}$, $x = -39.2° + 180n°.$
General solution is $x = 180n° \pm 39.2°.$

10 Parametric Equations

Expressions of the coordinates (or position vector) of a point on a curve in terms of a parameter.
Loci with simple equations in Cartesian and parametric forms.

10.1 Fact Sheet

(a) Parametric Form

Definition: When the equation of a curve $y = f(x)$ is expressed as $x = x(t)$, $y = y(t)$, these are the parametric equations of the curve. The independent variable t is the parameter.

(b) Gradient in Parametric Form

If $x = x(t)$, $y = y(t)$; then the gradient

$$\frac{dy}{dx} = \frac{dy}{dt} \div \frac{dx}{dt}$$

$$= \frac{dy}{dt} \frac{dt}{dx}.$$

(c) Circle

- $x^2 + y^2 = r^2 \Rightarrow x = r \cos t, \ y = r \sin t.$
- $(x - a)^2 + (y - b)^2 = r^2 \Rightarrow x = a + r \cos t, \ y = b + r \sin t.$

(d) Parabola

- $y^2 = 4ax \Rightarrow x = at^2, \ y = 2at.$

(e) Ellipse

- $\dfrac{x^2}{a^2} + \dfrac{y^2}{b^2} = 1 \Rightarrow x = a \cos t, \ y = b \sin t.$

(f) Rectangular Hyperbola

- $xy = c^2 \Rightarrow x = ct, \ y = \dfrac{c}{t}.$

10.2 Worked Examples

10.1 A curve is given parametrically by $x = t^2$, $y = t^3$. Find the equation of the tangent at the point with parameter t. If the tangent at P (p^2, p^3) meets the curve again at R find the coordinates of R and the equation of the normal at R.

- If $x = t^2$, then $\dfrac{dx}{dt} = 2t$, $y = t^3$, then $\dfrac{dy}{dt} = 3t^2$.

  But
  $$\frac{dy}{dx} = \left(\frac{dy}{dt}\right)\left(\frac{dt}{dx}\right) \quad \Rightarrow \quad \frac{dy}{dx} = \frac{3t^2}{2t} = \frac{3t}{2}. \tag{1}$$

  Therefore the equation of the tangent at (t^2, t^3) is

  $$y - t^3 = \frac{3t}{2}(x - t^2) \quad \Rightarrow \quad 2y - 2t^3 = 3tx - 3t^3 \quad \Rightarrow \quad 2y = 3tx - t^3.$$

  When the tangent at P, $2y = 3px - p^3$, meets the curve at R
  then $x = r^2$, $y = r^3$ $\Rightarrow$ $2r^3 = 3pr^2 - p^3$,

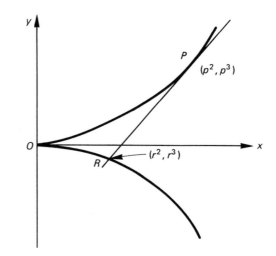

i.e. $\qquad 2r^3 - 3pr^2 + p^3 = 0$.

Two roots of this equation are $r = p$ (twice — because of the tangent at P).

Hence $\qquad 2r^3 - 3pr^2 + p^3 = 0 \quad \Rightarrow \quad (r - p)^2(2r + p) = 0$,

i.e. at the point R, $r = -p/2$ $\quad \Rightarrow \quad$ R is the point $\left(\dfrac{p^2}{4}, -\dfrac{p^3}{8}\right)$.

The parameter of R is $r = \dfrac{-p}{2}$.

Therefore at R $\dfrac{dy}{dx} = \dfrac{3r}{2} = \dfrac{-3p}{4}$.

The gradient of the normal at R is $\dfrac{-1}{-3p/4} = \dfrac{4}{3p}$.

The equation of the normal at R is $y - \left(-\dfrac{p^3}{8}\right) = \dfrac{4}{3p}\left(x - \dfrac{p^2}{4}\right)$.

i.e. $\qquad 24py = 32x - 8p^2 - 3p^4$.

10.2 A curve is given parametrically by the equations

$$x = \frac{1}{2}\left(t^2 + \frac{1}{t^2}\right), \quad y = \frac{1}{2}\left(t^2 - \frac{1}{t^2}\right), \qquad (t \neq 0).$$

(a) Find the Cartesian equation of the curve.
(b) Show that $x \geqslant 1$ for all t.
(c) Find the equation of the tangent at the point P with parameter p.

● (a) $$x = \frac{1}{2}\left(t^2 + \frac{1}{t^2}\right), \quad y = \frac{1}{2}\left(t^2 - \frac{1}{t^2}\right).$$

Adding gives $x + y = t^2$, subtracting gives $x - y = \frac{1}{t^2}$.

Multiplying: $(x + y)(x - y) = 1$.
Cartesian equation is $x^2 - y^2 = 1$.

(b) From (a), $x^2 = 1 + y^2$ and $x > 0$ from the parametric equation,
i.e. $x^2 \geqslant 1 \quad \Rightarrow \quad x \geqslant 1$ for all t.

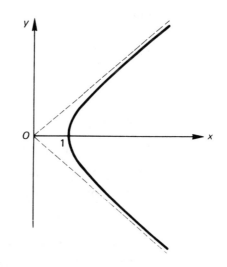

(c) From the Cartesian equation, $2x = 2y\, \dfrac{dy}{dx}$,

$$\frac{dy}{dx} = \frac{x}{y} = \left(t^2 + \frac{1}{t^2}\right) \div \left(t^2 - \frac{1}{t^2}\right) = \frac{t^4 + 1}{t^4 - 1}.$$

Equation of the tangent is

$$y - \frac{1}{2}\left(t^2 - \frac{1}{t^2}\right) = \frac{(t^4 + 1)}{(t^4 - 1)}\left[x - \frac{1}{2}\left(t^2 + \frac{1}{t^2}\right)\right].$$

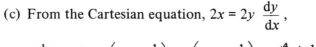

$$t^2(t^4 - 1)y - \tfrac{1}{2}(t^4 - 1)^2 = t^2(t^4 + 1)x - \tfrac{1}{2}(t^4 + 1)^2.$$

i.e. $(t^4 - 1)y = (t^4 + 1)x - 2t^2$.
At P, $t = p \quad \Rightarrow \quad$ tangent is $(p^4 - 1)y = (p^4 + 1)x - 2p^2$.

10.3 Show that the equation of the tangent to the parabola $y^2 = 4ax$ at the point $P(ap^2, 2ap)$ is $py = x + ap^2$.

If the chord PQ passes through the point $S(a, 0)$, prove that the tangents at P and Q intersect at right angles on a fixed straight line.

● At a general point on the parabola with parameter t,

$$x = at^2 \quad \Rightarrow \quad \frac{dx}{dt} = 2at, \qquad y = 2at, \quad \Rightarrow \quad \frac{dy}{dt} = 2a.$$

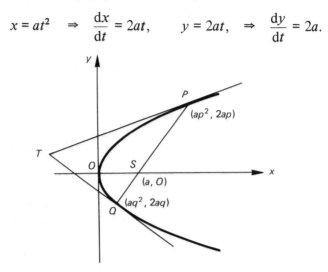

But
$$\frac{dy}{dx} = \left(\frac{dy}{dt}\right)\left(\frac{dt}{dx}\right) \quad \Rightarrow \quad \frac{dy}{dx} = \frac{2a}{2at} = \frac{1}{t}.$$

The equation of the tangent at $(at^2, 2at)$ is

$$y - 2at = \frac{1}{t}(x - at^2) \quad \Rightarrow \quad ty = x + at^2.$$

At P, $t = p$, tangent is $\qquad\qquad\qquad py = x + ap^2$. $\qquad$ (1)

At Q, $t = q$, tangent is $\qquad\qquad\qquad qy = x + aq^2$. $\qquad$ (2)

At the point of intersection equations 1 and 2 are satisfied simultaneously.
Equation 1 − equation 2 gives $py - qy = ap^2 - aq^2$.

$$y(p - q) = a(p - q)(p + q) \quad \Rightarrow \quad y = a(p + q) \qquad (p \neq q).$$

Substitute into equation 1: $ap(p + q) = x + ap^2$, so that $x = apq$.
Tangents at P and Q intersect at $T[apq, a(p + q)]$.
Since the chord passes through S, then the gradient of PS is equal to the gradient of SQ;

i.e. $\qquad\qquad \dfrac{2ap}{ap^2 - a} = \dfrac{2aq}{aq^2 - a} \quad \Rightarrow \quad \dfrac{p}{p^2 - 1} = \dfrac{q}{q^2 - 1}$

$\qquad \Rightarrow \quad p(q^2 - 1) = q(p^2 - 1) \quad \Rightarrow \quad pq^2 - qp^2 = p - q$

$\qquad \Rightarrow \quad pq(q - p) = p - q.$

Since $p \neq q$, divide by $(q - p)$, $\quad \Rightarrow \quad pq = -1$.

But the tangents at P and Q have gradients $\dfrac{1}{p}$ and $\dfrac{1}{q}$, respectively.

The product of the gradients $= \dfrac{1}{pq} = -1$.

Hence the tangents intersect at right angles.
The coordinates of T become $[-a, a(p + q)]$.
Hence T lies on the line $x = -a$ for all values of p and q.

10.4 The line $ty = x - 2t^2$ meets the curve $y = 2/x$ at the points $A(x_1, y_1)$ and $B(x_2, y_2)$. Form a quadratic equation in y with roots y_1 and y_2 and hence find the coordinates of the midpoint M of AB in terms of t.

Find the x, y equation of the locus of M.

Calculate

(a) the values of t for which A and B are real, distinct points.

(b) the value of t for which A and B are coincident.

Sketch the curve $xy = 2$, the line $ty = x - 2t^2$ using the value of t obtained in (b), and the locus of M.

●
$$ty = x - 2t^2, \tag{1}$$

$$y = 2/x, \qquad \text{hence } xy = 2. \tag{2}$$

The coordinates of A and B satisfy both of these equations.

From equation 1, $x = ty + 2t^2$. In equation 2, $y(ty + 2t^2) = 2$,

i.e.
$$ty^2 + 2t^2 y - 2 = 0. \tag{3}$$

Roots of this equation are y_1 and y_2 where $y_1 + y_2 = -2t$.

Midpoint M of AB is $\left(\dfrac{(x_1 + x_2)}{2}, \dfrac{(y_1 + y_2)}{2} \right)$.

Therefore at M, $y = -t$.

Substituting into equation 1: $-t^2 = x - 2t^2$, $x = t^2$.

Hence M has coordinates $(t^2, -t)$.

Parametric equation of locus of M is $x = t^2$, $y = -t$.

Cartesian equation is $x = y^2$.

(a) For A and B to be real distinct points, equation 3 must have real distinct roots, i.e. discriminant is positive,

i.e.
$$4t^4 + 8t > 0,$$

$$4t(t^3 + 2) > 0,$$

i.e.
$$t > 0 \quad \text{or} \quad t < (-2)^{1/3} = -(2)^{1/3}.$$

	$-(2)^{1/3}$		O	
$4t$	$-$		$-$	$+$
$t^3 + 2$	$-$		$+$	$+$
$4t(t^3 + 2)$	$+$		$-$	$+$

(b) When A and B are coincident, chord AB becomes a tangent and equation 3 has equal roots, i.e. discriminant = 0,

$$\Rightarrow \quad 4t^4 + 8t = 0, \qquad t = 0 \text{ or } (-2)^{1/3}.$$

When $t = 0$, equation 1 becomes $x = 0$, which is the y-axis. This is not a tangent for real finite values of y.

Therefore the line is a tangent when $t = (-2)^{1/3} = -2^{1/3}$.

Equation of tangent is $x + 2^{1/3} y = 2^{5/3}$.

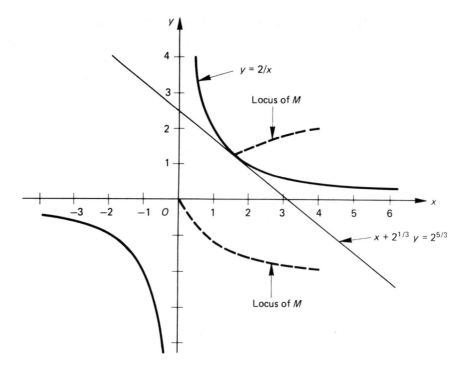

y = 2/x

Locus of M

Locus of M

$x + 2^{1/3} y = 2^{5/3}$

10.5 In each case eliminate the parameter t from the two equations to give an equation in x and y:

(a) $x = \dfrac{2t}{1 + t^3}$, $y = \dfrac{2t^2}{1 + t^3}$;

(b) $x = 3 + 2 \sec t$, $y = 4 - 5 \tan t$;

(c) $x = 4 \sin^2 t \cos t$, $y = 4 \sin t \cos^2 t$;

(d) $x = \dfrac{e^t + e^{-t}}{2}$, $y = \dfrac{e^t - e^{-t}}{2}$.

● (a) $x = \dfrac{2t}{1 + t^3}$, $y = \dfrac{2t^2}{1 + t^3}$.

Since x and y have a large common factor, find $\dfrac{x}{y}$ or $\dfrac{y}{x}$.

$\dfrac{y}{x} = t$. Substitute into the x equation:

$$x \left(1 + \frac{y^3}{x^3} \right) = \frac{2y}{x} .$$

Multiply by x^2 : $x^3 + y^3 = 2xy$.

(b) $x = 3 + 2 \sec t$; $y = 4 - 5 \tan t$.

Use the basic trigonometric identity $\sec^2 t = 1 + \tan^2 t$.

$$\sec t = \frac{x - 3}{2} , \qquad \tan t = \frac{4 - y}{5} ,$$

$$\Rightarrow \quad \frac{(x - 3)^2}{4} = 1 + \frac{(4 - y)^2}{25}$$

$$\Rightarrow \quad 25 (x - 3)^2 - 4 (4 - y)^2 = 100.$$

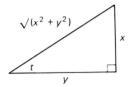

(c)

$$x = 4 \sin^2 t \cos t, \qquad (1)$$
$$y = 4 \sin t \cos^2 t, \qquad (2)$$

$$\therefore \frac{x}{y} = \frac{\sin t}{\cos t} = \tan t.$$

From the sketch, $\sin^2 t = \dfrac{x^2}{x^2 + y^2}$, $\cos^2 t = \dfrac{y^2}{x^2 + y^2}$.

From equations 1 and 2, $\quad x^2 + y^2 = 16 \sin^2 t \cos^2 t(\sin^2 t + \cos^2 t)$

$$= 16 \sin^2 t \cos^2 t$$
$$= \frac{16 x^2 y^2}{(x^2 + y^2)^2},$$
$$\Rightarrow \quad (x^2 + y^2)^3 = 16 x^2 y^2.$$

(d)

$$x = \frac{e^t + e^{-t}}{2}, \qquad (1)$$

$$y = \frac{e^t - e^{-t}}{2}. \qquad (2)$$

Add equations 1 and 2: $x + y = e^t$.
Subtract 2 from 1: $x - y \quad = e^{-t}$.
Multiply: $(x + y)(x - y) \quad = (e^t)(e^{-t}) = 1$
$$\Rightarrow \quad x^2 - y^2 = 1.$$

10.6 Prove that the equation of the normal at the point $\left(ct, \dfrac{c}{t} \right)$ on the rectangular hyperbola $xy = c^2$ is $\quad t^2 x - y - ct^3 + \dfrac{c}{t} = 0.$

The normal at a point P on the hyperbola $xy = c^2$ meets the x-axis at A, and the tangent at P meets the y-axis at B. Show that the locus of the midpoint of AB as P varies is $2c^2 xy = c^4 - y^4$.

● Using the parametric equation of the hyperbola,

$$x = ct, \quad \frac{dx}{dt} = c; \qquad y = \frac{c}{t}, \quad \frac{dy}{dt} = -\frac{c}{t^2}; \qquad \frac{dy}{dx} = \left(\frac{dy}{dt} \right) \left(\frac{dt}{dx} \right) = -\frac{1}{t^2}.$$

Gradient of the normal $= -\dfrac{dx}{dy} = t^2.$

Equation of the normal at P is

$$y - \frac{c}{t} = t^2 (x - ct) \quad \Rightarrow \quad t^2 x - y - ct^3 + \frac{c}{t} = 0.$$

At A, $y = 0$, $x = \dfrac{1}{t^2} \left(ct^3 - \dfrac{c}{t} \right) = \dfrac{c(t^4 - 1)}{t^3} \quad \Rightarrow \quad A\left(\dfrac{c(t^4 - 1)}{t^3}, 0 \right).$

Equation of the tangent at P is

$$y - \frac{c}{t} = -\frac{1}{t^2} (x - ct) \quad \Rightarrow \quad x + t^2 y = 2ct.$$

At B, $x = 0$, $y = \dfrac{2c}{t} \quad \Rightarrow \quad B\left(0, \dfrac{2c}{t} \right).$

109

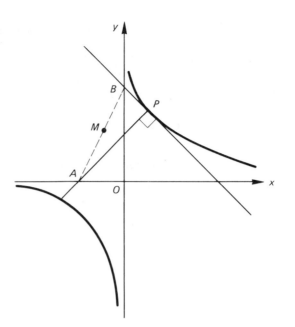

Midpoint of AB is M, $\left(\dfrac{c(t^4 - 1)}{2t^3} \, , \dfrac{c}{t} \right)$.

Parametric equations of the locus of M are

$$x = \frac{c(t^4 - 1)}{2t^3}, \tag{1}$$

$$y = \frac{c}{t}. \tag{2}$$

From equation 2, $t = \dfrac{c}{y}$.

Substitute into equation 1: $2x \, \dfrac{c^3}{y^3} = c\left(\dfrac{c^4}{y^4} - 1 \right)$.

Hence the locus of M is $2c^2 xy = c^4 - y^4$.

10.7 Find the equation of the tangent to the parabola $y^2 = 4ax$ at the point $P \, (ap^2 \, , 2ap)$. If the tangents at P and Q, with parameters p and q respectively, intersect at T, find the locus of T, given that PQ is of constant length l.

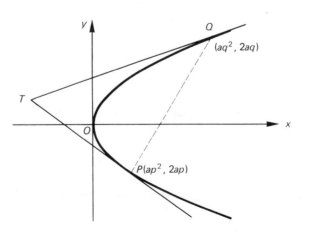

- Parametric equations of the parabola $x = at^2$, $y = 2at$:

$$\frac{dx}{dt} = 2at, \quad \frac{dy}{dt} = 2a, \quad \frac{dy}{dx} = \left(\frac{dy}{dt}\right)\left(\frac{dt}{dx}\right) = \frac{1}{t}.$$

Equation of the tangent is

$$y - 2at = \frac{1}{t}(x - at^2) \quad \Rightarrow \quad ty = x + at^2.$$

Tangent at P is $\qquad\qquad\qquad\qquad py = x + ap^2.$ $\qquad\qquad$ (1)
Tangent at Q is $\qquad\qquad\qquad\qquad qy = x + aq^2.$ $\qquad\qquad$ (2)
Solve simultaneously to find the coordinates of T.
Subtracting equation 2 from equation 1 gives
$$y(p - q) = a(p^2 - q^2) \qquad (p \neq q)$$
$$y = a(p + q).$$
Substitute into equation 1: $ap(p + q) = x + ap^2$
$$x = apq.$$
T has parametric equations $\qquad\qquad x = apq, \qquad y = a(p + q).$

(Length of PQ)^2 $= (ap^2 - aq^2)^2 + (2ap - 2aq)^2.$
$$l^2 = a^2(p^2 - q^2)^2 + 4a^2(p - q)^2$$
$$= a^2(p - q)^2[(p + q)^2 + 4].$$
But $(p - q)^2 = (p + q)^2 - 4pq.$
Therefore $\qquad\qquad l^2 = a^2[(p + q)^2 - 4pq][(p + q)^2 + 4].$ $\qquad$ (3)

From the parametric equations of T, $pq = \dfrac{x}{a}$, $(p + q) = \dfrac{y}{a}$.

Substitute into equation 3: $l^2 = a^2\left(\dfrac{y^2}{a^2} - \dfrac{4x}{a}\right)\left(\dfrac{y^2}{a^2} + 4\right).$

Multiply by a^2 : $a^2 l^2 = (y^2 - 4ax)(y^2 + 4a^2).$
Hence the locus of T is $a^2 l^2 = (y^2 - 4ax)(y^2 + 4a^2).$

10.3 Exercises

10.1
(a) Find the Cartesian equation of the curve with parametric equations
$$x = 3\cos t - 2, \qquad y = 5 - 2\sin t.$$

Sketch the curve for $0 \leqslant t \leqslant \dfrac{\pi}{2}$.

(b) Find the equation of the tangent to the curve given parametrically by

$$x = a(t + \cos t), \quad y = a(1 - \sin t), \text{ at the point where } t = \dfrac{\pi}{4}.$$

10.2 Find the equation of the tangent to the curve given parametrically by $x = 3t^2$, $y = 4t$. Find the point of intersection T of the tangents at points P and Q with parameters p and q respectively. If PQ subtends a right angle at the origin find the locus of T in Cartesian form.

10.3 Find the equation of the normal to the rectangular hyperbola at $P\left(cp, \dfrac{c}{p}\right)$.

If the normal meets the rectangular hyperbola again at $Q\left(cq, \dfrac{c}{q}\right)$ find q. Write down the equation of the circle on PQ as diameter. If this circle intersects the rectangular hyperbola again at R, find the parameter of R.

10.4 A curve is given parametrically by $x = \dfrac{t}{1 + t^3}$, $y = \dfrac{t^2}{1 + t^3}$.

Find (a) the Cartesian equation of the curve,
     (b) the equation of the tangent when $t = 3$.

10.5 In each case eliminate the parameter t from the two equations to give an equation in x and y:
(a) $x = 2t + 3$, $y = t^2 + 5t + 3$.
(b) $x = 4t^2 + 3t + 1$, $y = t^2$.
(c) $x = 3 + \sin^2 t$, $y = 4 - 2\cos^2 t$.
(d) $x = 2 + \sec^2 t$, $y = 3 + 2\tan^2 t$.

(e) $x = \dfrac{t}{3t^2 - 4}$, $y = \dfrac{2t^2}{3t^2 - 4}$.

Sketch the graphs of (a) and (d).

10.6 The tangents at $P\,(a\cos t,\, b\sin t)$ and $Q\,(a\cos z,\, b\sin z)$ to the ellipse $\dfrac{x^2}{a^2} + \dfrac{y^2}{b^2} = 1$ intersect at point T. Find the coordinates of T.

If $z = t + \dfrac{\pi}{2}$ prove that the locus of T is $\dfrac{x^2}{a^2} + \dfrac{y^2}{b^2} = 2$.

10.7 A curve is given parametrically by the equations

$$x = \frac{1}{2}\left(t + \frac{1}{t}\right), \quad y = \frac{1}{2}\left(t - \frac{1}{t}\right) \qquad (t \neq 0).$$

(a) Find the Cartesian equation of the curve.
(b) Find the equation of the normal to the curve at the point P with parameter p.
(c) Find the parameter of point R, where the normal intersects the curve again.

10.4   Outline Solutions to Exercises

10.1 (See diagram at the top of the next page).

(a) $\cos t = \dfrac{x + 2}{3}$, $\sin t = \dfrac{5 - y}{2}$.

    Equation $4(x + 2)^2 + 9(5 - y)^2 = 36$.
    Ellipse, centre $(-2, 5)$, major axis length 6, minor axis length 4.

(b) $\dfrac{dx}{dt} = a(1 - \sin t)$, $\dfrac{dy}{dt} = -a\cos t$ $\Rightarrow$ $\dfrac{dy}{dx} = \dfrac{-\cos t}{1 - \sin t}$.

When $t = \dfrac{\pi}{4}$, $x = a\left(\dfrac{\pi}{4} + \dfrac{\sqrt{2}}{2}\right)$, $y = a\left(1 - \dfrac{\sqrt{2}}{2}\right)$, $\dfrac{dy}{dx} = \dfrac{-1}{\sqrt{2} - 1}$.

    Tangent: $y - a\left(1 - \dfrac{\sqrt{2}}{2}\right) = \dfrac{-1}{\sqrt{2} - 1}\left[x - a\left(\dfrac{\pi}{4} + \dfrac{\sqrt{2}}{2}\right)\right]$

or $$y(\sqrt{2} - 1) = -x + a\left(\frac{\pi}{4} - 2 + 2\sqrt{2}\right).$$

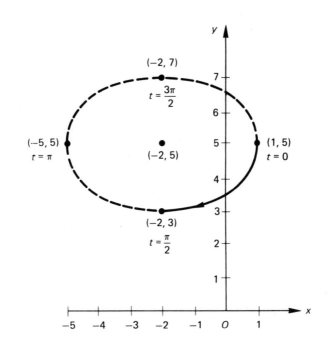

10.2 $\dfrac{\mathrm{d}y}{\mathrm{d}x} = \dfrac{2}{3t}$, tangent $3ty = 2x + 6t^2$.

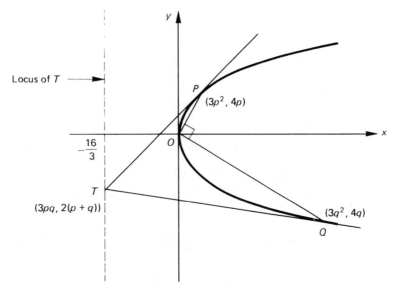

At P, $3py = 2x + 6p^2$.　At Q, $3qy = 2x + 6q^2$.
At T, $x = 3pq$, $y = 2(p + q)$.
Gradient $OP = \dfrac{4}{3p}$, gradient $OQ = \dfrac{4}{3q}$　$\Rightarrow$　$16 = -9pq$.

At T, $x = -\dfrac{16}{3}$, i.e. locus of T is the line $x = -\dfrac{16}{3}$.

10.3 Normal at P, $y = p^2 x - cp^3 + \dfrac{c}{p}$.

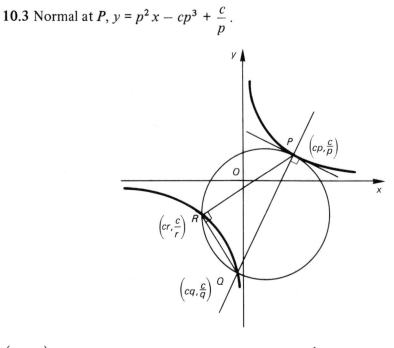

$\left(cq, \dfrac{c}{q}\right)$ satisfies equation of normal, $\Rightarrow$ $q = \dfrac{-1}{p^3}$.

Circle on PQ as diameter $(x - cp)(x - cq) + \left(y - \dfrac{c}{p}\right)\left(y - \dfrac{c}{q}\right) = 0$.

Passes through $\left(cr, \dfrac{c}{r}\right)$ $\Rightarrow$ $(r - p)(r - q) + \left(\dfrac{1}{r} - \dfrac{1}{p}\right)\left(\dfrac{1}{r} - \dfrac{1}{q}\right) = 0$.

$\Rightarrow$ $(r - p)(r - q)(r^2 pq + 1) = 0$.

$r \neq p, r \neq q$ $\Rightarrow$ $r^2 pq + 1 = 0$.

But $pq = \dfrac{-1}{p^2}$ $\Rightarrow$ $r = \pm p$, $r \neq p$ $\Rightarrow$ $r = -p$.

10.4

(a) $\dfrac{y}{x} = t$. Substitute in the x-equation: $x\left(1 + \dfrac{y^3}{x^3}\right) = \dfrac{y}{x}$.

Multiply by x^2: $x^3 + y^3 = xy$.

(b) $\dfrac{dx}{dt} = \dfrac{1 - 2t^3}{(1 + t^3)^2}$, $\dfrac{dy}{dt} = \dfrac{t(2 - t^3)}{(1 + t^3)^2}$, $\dfrac{dy}{dx} = \dfrac{t(2 - t^3)}{(1 - 2t^3)}$.

When $t = 3$, $x = \frac{3}{28}$, $y = \frac{9}{28}$, $\dfrac{dy}{dx} = \frac{75}{53}$.

Tangent is $y - \frac{9}{28} = \frac{75}{53}(x - \frac{3}{28})$ or $53y = 75x + 9$.

10.5

(a) $t = \dfrac{x - 3}{2}$, $y = \dfrac{(x - 3)^2}{4} + \dfrac{5(x - 3)}{2} + 3$

$4y = x^2 + 4x - 9$. (See the diagram at the top of the next page.)

(b) $x - 4y = 3t + 1$ $\Rightarrow$ $(x - 4y - 1)^2 = 9t^2 = 9y$.

(c) $x - 3 = \sin^2 t$, $\dfrac{y - 4}{2} = -\cos^2 t$ $\Rightarrow$ $2(x - 3) - (y - 4) = 2$

$\Rightarrow$ $2x - y = 4$.

Line segment: $3 \leqslant x \leqslant 4$, $2 \leqslant y \leqslant 4$.

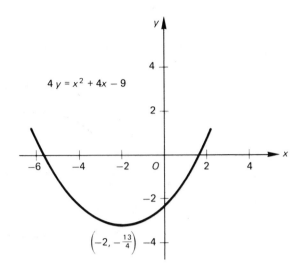

$4y = x^2 + 4x - 9$

$\left(-2, -\frac{13}{4}\right)$

(d) $\sec^2 t = x - 2$, $\quad 2\tan^2 t = y - 3$, $\quad \Rightarrow \quad 2(x - 2) = y - 3 + 2$
$$\Rightarrow \quad 2x - y = 3.$$

Half line: $x \geqslant 3, y \geqslant 3$.

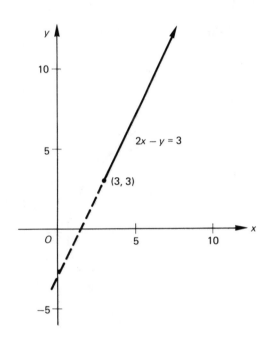

$2x - y = 3$

$(3, 3)$

(e) $\dfrac{y}{x} = 2t$, $\dfrac{y^2}{x^2} = 4t^2$, but $t^2 = \dfrac{4y}{3y - 2}$ $\quad \Rightarrow \quad 3y^2 - 2y = 16x^2$.

10.6 Tangent at P is $\dfrac{x \cos t}{a} + \dfrac{y \sin t}{b} = 1$, at Q is $\dfrac{x \cos z}{a} + \dfrac{y \sin z}{b} = 1$.

Intersection when these are solved simultaneously:
$$x = \frac{a(\sin z - \sin t)}{\sin(z - t)}, \ y = \frac{b(\cos t - \cos z)}{\sin(z - t)}.$$

If $z = t + \dfrac{\pi}{2}$, $x = a(\cos t - \sin t)$, $y = b(\cos t + \sin t)$.

$$\frac{x^2}{a^2} + \frac{y^2}{b^2} = (\cos t - \sin t)^2 + (\cos t + \sin t)^2 = 2.$$

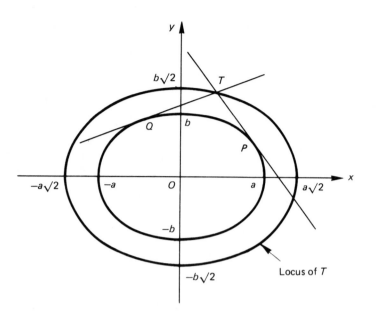

10.7

(a) Add: $x + y = t$. Subtract: $x - y = \dfrac{1}{t}$. Multiply: $x^2 - y^2 = 1$.

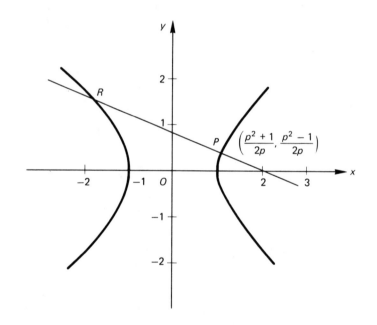

(b) $\dfrac{dy}{dx} = \dfrac{t^2 + 1}{t^2 - 1}$.

    At P, $y = \dfrac{p^2 - 1}{2p}$, $x = \dfrac{p^2 + 1}{2p}$, $\dfrac{dy}{dx} = \dfrac{p^2 + 1}{p^2 - 1}$.

    Normal is $2p(p^2 + 1)y - (p^2 - 1)(p^2 + 1) = -2p(p^2 - 1)x + p^4 - 1$
               $\Rightarrow$ $2p(p^2 + 1)y + 2p(p^2 - 1)x = 2p^4 - 2$.

    Meets the curve again at $R\left(\dfrac{r^2 + 1}{2r}, \dfrac{r^2 - 1}{2r}\right)$;

           $2p(p^2 + 1)(r^2 - 1) + 2p(p^2 - 1)(r^2 + 1) = 2r(2p^4 - 2)$
               $\Rightarrow$ $p^3 r^2 + r(1 - p^4) - p = 0$

This is a quadratic in r. One root is p.

Product of roots is $\dfrac{-1}{p^2}$, $\quad\Rightarrow\quad$ second root is $\dfrac{-1}{p^3}$.

Parameter of R is $-\dfrac{1}{p^3}$.

11 Vectors

Vectors in two and three dimensions. Algebraic operations of addition and multiplication by scalars and their geometrical significance. The scalar product and its use for calculating the angle between two coplanar lines. Position vectors; the vector equation of a line in the form $\mathbf{r} = \mathbf{a} + t\mathbf{b}$ and $\mathbf{r} = t\mathbf{a} + (1 - t)\mathbf{b}$.

11.1 Fact Sheet

(a) Algebra

- A vector $\mathbf{a}$ has magnitude or modulus $|\mathbf{a}| = a$ and direction $\hat{\mathbf{a}}$ (where $\hat{\mathbf{a}}$ is a unit vector): $\mathbf{a} = a\hat{\mathbf{a}}$ or $\hat{\mathbf{a}} = \dfrac{\mathbf{a}}{a}$.

- If λ is any positive scalar then $\lambda\mathbf{a}$ has magnitude λa and direction $\hat{\mathbf{a}}$. If λ is negative then $\lambda\mathbf{a}$ has magnitude $|\lambda|a$ and direction $-\hat{\mathbf{a}}$, i.e. in the opposite direction.

(b) Position Vectors

If O is an origin then $\overline{OA}$ is the position vector of A relative to O. The usual notation is $\overline{OA} = \mathbf{a}$, $\overline{OB} = \mathbf{b}$ etc.

(c) Displacement Vectors

If $\overline{OA} = \mathbf{a}$ and $\overline{OB} = \mathbf{b}$ then $\overline{AB}$ is the displacement vector from A to B, and is given by $\overline{AB} = \mathbf{b} - \mathbf{a}$.

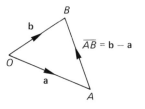

(d) Ratio Theorem

- If $\overline{OA} = \mathbf{a}$ and $\overline{OB} = \mathbf{b}$, then the position vector of the point P which divides $\overline{AB}$ in the ratio $m : n$ is $\overline{OP} = \mathbf{p} = \dfrac{n\mathbf{a} + m\mathbf{b}}{m + n}$.

- If m and n are both positive, P_1 lies between A and B. If m is positive and n is negative, P_2 lies on AB produced. If m is negative and n is positive, P_3 lies on BA produced.

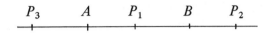

(e) Equations of Lines

If a line is drawn through a point A, parallel to the vector $\mathbf{b}$, the position vector of any point P on the line is given by $\mathbf{r} = \mathbf{a} + t\mathbf{b}$, where t is a parameter.

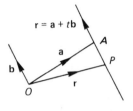

From the ratio theorem, putting $\dfrac{m}{m + n} = 1 - t$ and $\dfrac{n}{m + n} = t$, the position vector of any point on the line AB is given by $\mathbf{r} = t\mathbf{a} + (1 - t)\mathbf{b}$.

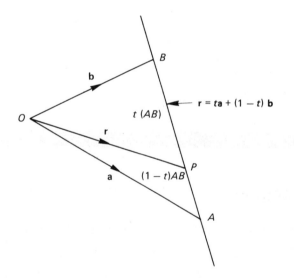

(f) Cartesian Components

$\mathbf{i}$, $\mathbf{j}$ and $\mathbf{k}$ are unit vectors in the positive directions of the x, y and z axes, and are called base vectors. A point P with coordinates (x, y, z) has a position vector

$$\mathbf{r} = x\mathbf{i} + y\mathbf{j} + z\mathbf{k} \text{ or } \mathbf{r} = \begin{pmatrix} x \\ y \\ z \end{pmatrix} \text{ or occasionally } (x, y, z) \text{ (see Worked Example 8).}$$

(g) Magnitude or Modulus

The magnitude or modulus of the vector $\mathbf{r} = x\mathbf{i} + y\mathbf{j} + z\mathbf{k}$ is $\sqrt{(x^2 + y^2 + z^2)}$.

(h) Scalar Product

The scalar product of the vectors **a** and **b** is written **a · b** and defined
a · b = |**a**| |**b**| $\cos \theta$ = $ab \cos \theta$ where θ is the angle between **a** and **b**.

(i) If **a** and **b** are parallel then **a · b** = ab.

(ii) If **a** = **b** then **a · b** = **a · a** = a^2.

(iii) If **a** and **b** are perpendicular then **a · b** = 0.

(iv) If **a** and **b** are given in Cartesian form, **a** = $a_1\mathbf{i} + a_2\mathbf{j} + a_3\mathbf{k}$, **b** = $b_1\mathbf{i} + b_2\mathbf{j} + b_3\mathbf{k}$,
then **a · b** = $a_1b_1 + a_2b_2 + a_3b_3$.

(v) If the angle between the vectors **a** and **b** is θ then

$$\cos \theta = \frac{a_1b_1 + a_2b_2 + a_3b_3}{ab} = \frac{\mathbf{a \cdot b}}{ab}.$$

(i) Projection

The projection of **a** on **b** is the resolved component (or resolute) of the vector **a** in
the direction of **b** and equals **a** · $\hat{\mathbf{b}}$ = $a \cos \theta$.

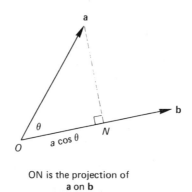

ON is the projection of
a on **b**

11.2 Worked Examples

11.1 Find a unit vector which is perpendicular to the vector $4\mathbf{i} + 4\mathbf{j} - 7\mathbf{k}$, and
to the vector $3\mathbf{i} - 2\mathbf{j} + \mathbf{k}$.

● Let a vector which is perpendicular to the given two vectors be $p\mathbf{i} + q\mathbf{j} + r\mathbf{k}$.

Then
$$\begin{pmatrix} 4 \\ 4 \\ -7 \end{pmatrix} \cdot \begin{pmatrix} p \\ q \\ r \end{pmatrix} = 0 \text{ and } \begin{pmatrix} 3 \\ -2 \\ 1 \end{pmatrix} \cdot \begin{pmatrix} p \\ q \\ r \end{pmatrix} = 0.$$

$$\Rightarrow \quad 4p + 4q - 7r = 0 \tag{1}$$

and

$$3p - 2q + r = 0. \tag{2}$$

Solving equations 1 and 2 simultaneously,

$$4p + 4q - 7r = 0$$
$$6p - 4q + 2r = 0.$$

Adding: $\qquad\qquad\qquad\qquad 10p - 5r = 0 \quad \Rightarrow \quad r = 2p.$

Substituting into equation 2:

$$3p - 2q + 2p = 0 \quad \Rightarrow \quad q = \tfrac{5}{2}p.$$

Any value of p may now be chosen.

Taking $p = 2$, we have $2\mathbf{i} + 5\mathbf{j} + 4\mathbf{k}$ is a vector perpendicular to the two given vectors.

Thus the required unit vector is

$$\frac{2\mathbf{i} + 5\mathbf{j} + 4\mathbf{k}}{\sqrt{(2^2 + 5^2 + 4^2)}} = \frac{2\mathbf{i} + 5\mathbf{j} + 4\mathbf{k}}{\sqrt{(45)}}.$$

11.2 Find the unit vector $\hat{\mathbf{a}}$ parallel to the vector $\mathbf{a} = 3\mathbf{i} - 2\mathbf{j} + \mathbf{k}$. Determine the length of the resolved part of the vector $\mathbf{b} = 3\mathbf{i} + \mathbf{j} + 2\mathbf{k}$ in the direction of $\mathbf{a}$.

●
$$\hat{\mathbf{a}} = \frac{\mathbf{a}}{a} = \frac{3\mathbf{i} - 2\mathbf{i} + \mathbf{k}}{\sqrt{(3^2 + 2^2 + 1^2)}} = \frac{1}{\sqrt{(14)}}(3\mathbf{i} - 2\mathbf{j} + \mathbf{k}).$$

$\mathbf{b} \cdot \hat{\mathbf{a}} =$ length of resolved part of $\mathbf{b}$ in the direction of $\mathbf{a}$

$$= \begin{pmatrix} 3 \\ 1 \\ 2 \end{pmatrix} \cdot \frac{1}{\sqrt{(14)}} \begin{pmatrix} 3 \\ -2 \\ 1 \end{pmatrix}$$

$$= \frac{1}{\sqrt{(14)}}(9 - 2 + 2) = \frac{9}{\sqrt{(14)}}$$

$\Rightarrow$ resolved part of $\mathbf{b}$ has magnitude $\dfrac{9}{\sqrt{(14)}}$.

11.3 $\mathbf{a} = \begin{pmatrix} \frac{2}{11} \\ p \\ -\frac{9}{11} \end{pmatrix}$ is a unit vector. Which of the following could be the value of p?

(i) $-\frac{4}{11}$;　　(ii) $\frac{6}{11}$;　　(iii) $\frac{18}{11}$;　　(iv) $-\frac{6}{11}$.

A, (i) only;　　　　　　　B, (i) and (iv) only;　　C, (iii) only;

D, (ii) and (iv) only;　　E, none of these.

● The magnitude of $\mathbf{a}$ is $\sqrt{[(\tfrac{2}{11})^2 + p^2 + (-\tfrac{9}{11})^2]}$

$$= \sqrt{(\tfrac{4}{121} + p^2 + \tfrac{81}{121})}.$$

But $\mathbf{a}$ is a unit vector so the magnitude of $\mathbf{a}$ is 1.

Therefore $1 = \tfrac{85}{121} + p^2$.

$p^2 = \tfrac{36}{121}, p = \pm\tfrac{6}{11}$. 　　　　　　　　　　　　　　　　Answer D

11.4 Which of the following values of x and y make the vectors $\begin{pmatrix} x \\ y \\ 5 \end{pmatrix}$ and $\begin{pmatrix} 3 \\ 4 \\ 2 \end{pmatrix}$ perpendicular?

(i) $x = -2, y = -1$;　　(ii) $x = 0, y = \tfrac{5}{2}$;　　(iii) $x = -6, y = 2$.

A, (i) only;　　　　　　　　B, (ii) only;　　C, (i) and (iii) only;

D, (i), (ii) and (iii);　　　E, (ii) and (iii) only.

● If the vectors are perpendicular then their scalar product is zero.

$$\begin{pmatrix} x \\ y \\ 5 \end{pmatrix} \cdot \begin{pmatrix} 3 \\ 4 \\ 2 \end{pmatrix} = 3x + 4y + 10 = 0.$$

Test the values given in (i), (ii) and (iii):
  (i) $3x + 4y + 10 = -6 - 4 + 10 = 0$; (i) is correct.
  (ii) $3x + 4y + 10 = 0 + 10 + 10 = 20$; (ii) is incorrect.
  (iii) $3x + 4y + 10 = -18 + 8 + 10 = 0$; (iii) is correct.

Answer **C**

11.5 $ABCD$ is a trapezium where AB is parallel to DC. If $\overline{AB}$ represents $\begin{pmatrix} 2 \\ 5 \\ 3 \end{pmatrix}$ and $\overline{DC}$ represents $\begin{pmatrix} p \\ q \\ r \end{pmatrix}$, which of the following must be true?

(i) $r = 3$; (ii) $p : r = 2 : 3$; (iii) $2p + 5q + 3r = 0$; (iv) $2p + 5q + 3r = 1$.
A, (i) only;   B, (ii) only;   C, (iii) only;
D, (ii) and (iv) only;   E, none of these.

● If $\overline{AB}$ is parallel to $\overline{DC}$, $\dfrac{2}{p} = \dfrac{5}{q} = \dfrac{3}{r} \Rightarrow p : r = 2 : 3$.

Answer **B**

11.6 The non-zero vectors **p**, **q** and **r** are such that **p** + **q** and **p** − **q** are respectively perpendicular to **p** + **r** and **p** − **r**. Prove that **p** is perpendicular to **q** + **r**.
   If the ratios of the magnitudes of the vectors **p**, **q**, **r** are respectively $1 : 2 : 3$, find the angle θ ($0 \leqslant \theta \leqslant \pi$) between the vectors **q** and **r**.

● If **p** + **q** is perpendicular to **p** + **r**,

$$(\mathbf{p} + \mathbf{q}) \cdot (\mathbf{p} + \mathbf{r}) = 0$$
$$\Rightarrow p^2 + \mathbf{q} \cdot \mathbf{p} + \mathbf{p} \cdot \mathbf{r} + \mathbf{q} \cdot \mathbf{r} = 0. \tag{1}$$

Similarly,

$$p^2 - \mathbf{q} \cdot \mathbf{p} - \mathbf{p} \cdot \mathbf{r} + \mathbf{q} \cdot \mathbf{r} = 0. \tag{2}$$

Subtract equation 2 from equation 1:

$$2(\mathbf{q} \cdot \mathbf{p} + \mathbf{p} \cdot \mathbf{r}) = 0 \Rightarrow \mathbf{p} \cdot (\mathbf{q} + \mathbf{r}) = 0$$
$$\Rightarrow \mathbf{p} \text{ is perpendicular to } \mathbf{q} + \mathbf{r}.$$

Adding equations 1 and 2 gives:

$$2(p^2 + \mathbf{q} \cdot \mathbf{r}) = 0 \tag{3}$$

If $p : q : r = 1 : 2 : 3$ then let $p = k$, $q = 2k$ and $r = 3k$.
Equation 3 gives $k^2 + (2k)(3k) \cos \theta = 0$.
Thus $\cos \theta = -\frac{1}{6}$, $\theta = 1.74$ radians.
Hence the angle between **q** and **r** is 1.74 radians.

11.7 In triangle ABC, E lies on BC with $BE/EC = \frac{2}{3}$, F lies on CA with $CF/FA = \frac{3}{4}$, and G lies on AB produced, with $GB/GA = \frac{1}{2}$. The position vectors of A, B and C relative to an origin O are **a**, **b** and **c** respectively. Determine the position vectors of E, F and G in terms of **a**, **b** and **c** and deduce that E, F and G lie on a straight line.

● If $BE/EC = \frac{2}{3}$ then $\overline{BE} = \frac{2}{5}\overline{BC}$, $\overline{EC} = \frac{3}{5}\overline{BC}$.
   Position vector of E, $\mathbf{e} = \frac{3}{5}\mathbf{b} + \frac{2}{5}\mathbf{c}$ (by the ratio theorem).

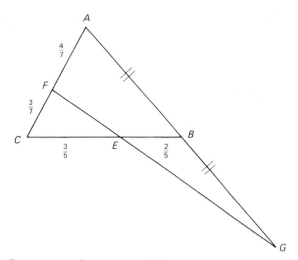

$CF/FA = \frac{3}{4}$, so $\overline{CF} = \frac{3}{7}\overline{CA}$, $\overline{FA} = \frac{4}{7}\overline{CA}$.

Position vector of F, $\mathbf{f} = \frac{4}{7}\mathbf{c} + \frac{3}{7}\mathbf{a}$.

$GB/GA = \frac{1}{2}$, so $BG/GA = -\frac{1}{2}$ $\Rightarrow$ $\overline{BG} = -\overline{BA}$, $\overline{GA} = 2\overline{BA}$.

Position vector of G, $\mathbf{g} = 2\mathbf{b} - \mathbf{a}$.

Displacement vector $\overline{FE} = \mathbf{e} - \mathbf{f} = \frac{3}{5}\mathbf{b} + \frac{2}{5}\mathbf{c} - \frac{4}{7}\mathbf{c} - \frac{3}{7}\mathbf{a}$

$\qquad\qquad\qquad\qquad = -\frac{3}{7}\mathbf{a} + \frac{3}{5}\mathbf{b} - \frac{6}{35}\mathbf{c}$

$\qquad\qquad\qquad\qquad = -\frac{3}{7}(\mathbf{a} - \frac{7}{5}\mathbf{b} + \frac{2}{5}\mathbf{c})$.

Displacement vector $\overline{EG} = \mathbf{g} - \mathbf{e} = 2\mathbf{b} - \mathbf{a} - \frac{3}{5}\mathbf{b} - \frac{2}{5}\mathbf{c}$

$\qquad\qquad\qquad\qquad = -\mathbf{a} + \frac{7}{5}\mathbf{b} - \frac{2}{5}\mathbf{c}$

$\qquad\qquad\qquad\qquad = -(\mathbf{a} - \frac{7}{5}\mathbf{b} + \frac{2}{5}\mathbf{c})$.

Therefore $\qquad\qquad \overline{FE} = \frac{3}{7}\overline{EG}$.

Hence E, F, G lie on a straight line.

11.8 The lines l_1 and l_2 are given in the parametric forms:

$$\mathbf{r}_1 = \mathbf{a}_1 + t_1\mathbf{b}_1, \qquad \text{where } a_1 = (3, 2, 1), \quad b_1 = (1, 2, 2);$$

$$\mathbf{r}_2 = \mathbf{a}_2 + t_2\mathbf{b}_2, \qquad \text{where } a_2 = (2, 3, 2), \quad b_2 = (2, 1, -2).$$

Show that l_1 and l_2 are perpendicular.

Find the values of t_1 and t_2 if the vector $\mathbf{r}_2 - \mathbf{r}_1$ is perpendicular to each of the given lines.

Hence find the length of the common perpendicular of l_1 and l_2.

● $\mathbf{r}_1 = \mathbf{a}_1 + t_1\mathbf{b}_1$ represents line l_1, with direction $\mathbf{b}_1$.

Similarly, direction of l_2 is given by $\mathbf{b}_2$.

$\mathbf{b}_1 \cdot \mathbf{b}_2 = (1, 2, 2) \cdot (2, 1, -2) = (1)(2) + (2)(1) + (2)(-2) = 0$.

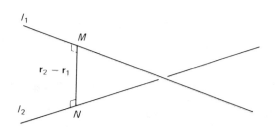

Hence $\mathbf{b}_1$ is perpendicular to $\mathbf{b}_2$, and so l_1 is perpendicular to l_2.

If $\mathbf{r}_2 - \mathbf{r}_1$ is perpendicular to l_1 then $(\mathbf{r}_2 - \mathbf{r}_1) \cdot \mathbf{b}_1 = 0$

$\Rightarrow \quad \mathbf{r}_2 \cdot \mathbf{b}_1 = \mathbf{r}_1 \cdot \mathbf{b}_1$

$\Rightarrow \quad (\mathbf{a}_2 + t_2\mathbf{b}_2) \cdot \mathbf{b}_1 = (\mathbf{a}_1 + t_1\mathbf{b}_1) \cdot \mathbf{b}_1$

$$\Rightarrow \quad \mathbf{a}_2 \cdot \mathbf{b}_1 + t_2 \mathbf{b}_2 \cdot \mathbf{b}_1 = \mathbf{a}_1 \cdot \mathbf{b}_1 + t_1 \mathbf{b}_1 \cdot \mathbf{b}_1.$$

But $\mathbf{b}_2 \cdot \mathbf{b}_1 = 0 \quad \Rightarrow \quad \mathbf{b}_1 \cdot (\mathbf{a}_2 - \mathbf{a}_1) = t_1 b_1^2$

$$\Rightarrow \quad (1, 2, 2) \cdot (-1, 1, 1) = t_1 \, (1^2 + 2^2 + 2^2)$$

$-1 + 2 + 2 = 9t_1 \quad \Rightarrow \quad t_1 = \frac{1}{3}.$

Similarly, $(\mathbf{r}_2 - \mathbf{r}_1) \cdot \mathbf{b}_2 = 0.$

$\Rightarrow \quad \mathbf{r}_2 \cdot \mathbf{b}_2 = \mathbf{r}_1 \cdot \mathbf{b}_2$

$\Rightarrow \quad (\mathbf{a}_2 + t_2 \mathbf{b}_2) \cdot \mathbf{b}_2 = (\mathbf{a}_1 + t_1 \mathbf{b}_1) \cdot \mathbf{b}_2$

$\Rightarrow \quad \mathbf{a}_2 \cdot \mathbf{b}_2 + t_2 \mathbf{b}_2 \cdot \mathbf{b}_2 = \mathbf{a}_1 \cdot \mathbf{b}_2 + t_1 \mathbf{b}_1 \cdot \mathbf{b}_2.$

But $\mathbf{b}_1 \cdot \mathbf{b}_2 = 0 \quad \Rightarrow \quad \mathbf{b}_2 \cdot (\mathbf{a}_1 - \mathbf{a}_2) = t_2 b_2^2.$

$$\Rightarrow \quad (2, 1, -2) \cdot (1, -1, -1) = t_2 \, (2^2 + 1^2 + (-2)^2).$$

$9t_2 = 2 - 1 + 2 = 3 \quad \Rightarrow \quad t_2 = \frac{1}{3}.$

Point M on l_1 is given by

$$\mathbf{r}_1 = (3, 2, 1) + \tfrac{1}{3}(1, 2, 2) = \tfrac{1}{3}(10, 8, 5).$$

Point N on l_2 is given by

$$\mathbf{r}_2 = (2, 3, 2) + \tfrac{1}{3}(2, 1, -2) = \tfrac{1}{3}(8, 10, 4),$$

and displacement vector $\mathbf{r}_2 - \mathbf{r}_1 = \tfrac{1}{3}(-2, 2, -1).$

Magnitude of displacement vector

$$|\mathbf{r}_2 - \mathbf{r}_1| = \sqrt{\{(-\tfrac{2}{3})^2 + (\tfrac{2}{3})^2 + (-\tfrac{1}{3})^2\}} = 1.$$

Hence the length of the common perpendicular is 1.

11.9 The point A (1, 4, 3) has position vector $\mathbf{a}$ relative to the origin O, and the point B (2, 1, 3) has position vector $\mathbf{b}$. The line l_1 is given by $\mathbf{r} = \mathbf{a} + s(2\mathbf{i} - \mathbf{j} + \mathbf{k})$; the line l_2 is given by $\mathbf{r} = \mathbf{b} + t(3\mathbf{i} + \mathbf{j} + 2\mathbf{k})$. Show that the lines l_1 and l_2 intersect and state the coordinates of the common point. Prove that AB is perpendicular to l_2.

● Two lines intersect if there exist values of s and t which will give the same position vector $\mathbf{r}$ for lines l_1 and l_2.

$$\text{For a point on line } l_1, \quad \mathbf{r}_1 = \begin{pmatrix} 1 + 2s \\ 4 - s \\ 3 + s \end{pmatrix}.$$

$$\text{For a point on line } l_2, \quad \mathbf{r}_2 = \begin{pmatrix} 2 + 3t \\ 1 + t \\ 3 + 2t \end{pmatrix}.$$

$$\text{If } \mathbf{r}_1 = \mathbf{r}_2, \quad 1 + 2s = 2 + 3t; \tag{1}$$
$$4 - s = 1 + t; \tag{2}$$
$$3 + s = 3 + 2t. \tag{3}$$

All three equations must be satisfied if the lines are to intersect.

Solving equations 2 and 3 by adding:

$$7 = 4 + 3t \quad \Rightarrow \quad t = 1.$$

Substitute into equation 3: $3 + s = 3 + 2, \quad \Rightarrow \quad s = 2.$

Check in equation 1: l.h.s., $1 + 2s = 5$; r.h.s., $2 + 3t = 5.$

Therefore $\mathbf{r}_1 = \mathbf{r}_2$ when $s = 2$ and $t = 1.$

The lines l_1 and l_2 intersect at (5, 2, 5).

Displacement vector $\overline{AB} = \mathbf{b} - \mathbf{a} = \begin{pmatrix} 2 \\ 1 \\ 3 \end{pmatrix} - \begin{pmatrix} 1 \\ 4 \\ 3 \end{pmatrix} = \begin{pmatrix} 1 \\ -3 \\ 0 \end{pmatrix}.$

$\overline{AB}$ is perpendicular to l_2 if $\overline{AB}$ is perpendicular to $3\mathbf{i} + \mathbf{j} + 2\mathbf{k}$, i.e. if the scalar product is zero.

Scalar product $= (1)(3) - (3)(1) + (0)(2) = 0$.

Hence AB is perpendicular to l_2.

11.10

(a) Use the scalar product to find the angle (to the nearest degree) between $3\mathbf{i} - \mathbf{j} + 2\mathbf{k}$ and $2\mathbf{i} + \mathbf{j} - \mathbf{k}$.

(b) The three vectors $3\mathbf{i} - \mathbf{j} + 2\mathbf{k}$, $\mathbf{i} + a\mathbf{j} - 2\mathbf{k}$ and $\mathbf{i} + b\mathbf{j} + c\mathbf{k}$ are mutually perpendicular; find a, b and c.

● (a) Scalar product $\mathbf{u} \cdot \mathbf{v} = uv \cos \theta$, where θ is the angle between $\mathbf{u}$ and $\mathbf{v}$.

$(3\mathbf{i} - \mathbf{j} + 2\mathbf{k}) \cdot (2\mathbf{i} + \mathbf{j} - \mathbf{k}) = 6 - 1 - 2 = 3$.

$u = $ magnitude of $\mathbf{u} = \sqrt{(3^2 + 1^2 + 2^2)} = \sqrt{14}$.

$v = $ magnitude of $\mathbf{v} = \sqrt{(2^2 + 1^2 + 1^2)} = \sqrt{6}$.

Therefore $\cos \theta = \dfrac{\mathbf{u} \cdot \mathbf{v}}{uv} = \dfrac{3}{\sqrt{(14)}\sqrt{6}} \quad \Rightarrow \quad \theta = 71°$.

(b) If $\mathbf{u}$ and $\mathbf{v}$ are perpendicular, $\mathbf{u} \cdot \mathbf{v} = 0$.

$(3\mathbf{i} - \mathbf{j} + 2\mathbf{k}) \cdot (\mathbf{i} + a\mathbf{j} - 2\mathbf{k}) = 0$.

$\Rightarrow \quad 3 - a - 4 = 0 \quad \Rightarrow \quad a = -1$.

$(3\mathbf{i} - \mathbf{j} + 2\mathbf{k}) \cdot (\mathbf{i} + b\mathbf{j} + c\mathbf{k}) = 0 \quad \Rightarrow \quad 3 - b + 2c = 0$, $\qquad$ (1)

$(\mathbf{i} + a\mathbf{j} - 2\mathbf{k}) \cdot (\mathbf{i} + b\mathbf{j} + c\mathbf{k}) = 0 \quad \Rightarrow \quad 1 + ab - 2c = 0$. $\qquad$ (2)

Put $a = -1$ into equation 2 and solve giving

$$a = -1, b = 2, c = -\tfrac{1}{2}.$$

11.3 Exercises

11.1 If $\mathbf{i}$, $\mathbf{j}$ and $\mathbf{k}$ are mutually perpendicular vectors, and

$$\mathbf{A} = \mathbf{i} + 2\mathbf{j} + 3\mathbf{k}, \qquad \mathbf{B} = 3\mathbf{i} + \mathbf{k}, \qquad \mathbf{C} = -2\mathbf{i} + 3\mathbf{j},$$

Calculate $|\mathbf{A} + 2\mathbf{B} - \mathbf{C}|$ and $\mathbf{B} \cdot (\mathbf{A} - \mathbf{C})$. Find a vector of length $2\sqrt{26}$ perpendicular to $\mathbf{A}$ and $\mathbf{B}$.

11.2 Given that A and B have position vectors

$$\mathbf{a} = \mathbf{i} + 2\mathbf{j} + 3\mathbf{k} \qquad \text{and} \qquad \mathbf{b} = -2\mathbf{i} + 5\mathbf{j} - \mathbf{k}$$

respectively, determine $\overline{AB}$, $|\overline{AB}|$ and the direction ratios of the line AB.

11.3 The resolved part of the force $\begin{pmatrix} 2a \\ 6a \\ -6a \end{pmatrix}$ N in the direction of $\begin{pmatrix} 2 \\ -1 \\ 2 \end{pmatrix}$ is 14 N.

Find the value of a.

11.4 The non-zero vectors $\mathbf{p}$, $\mathbf{q}$ and $\mathbf{r}$ have equal magnitude k and make angles $\pi/3$ with each other. Prove that the vectors $\mathbf{p} + \mathbf{q} - \mathbf{r}$, $\mathbf{p} - \mathbf{q} + \mathbf{r}$ and $-\mathbf{p} + \mathbf{q} + \mathbf{r}$ are mutually perpendicular and have magnitude $k\sqrt{2}$.

OP, OQ and OR are three concurrent edges of a parallelepiped, given respectively by the vectors $\mathbf{p}$, $\mathbf{q}$ and $\mathbf{r}$. Interpret the above result in terms of the parallelepiped.

11.5 The lines l_1 and l_2 are given by

$$\mathbf{r}_1 = \begin{pmatrix} 3 \\ 1 \\ 4 \end{pmatrix} + s \begin{pmatrix} 5 \\ -2 \\ 3 \end{pmatrix} \quad \text{and} \quad r_2 = \begin{pmatrix} -3 \\ 5 \\ 2 \end{pmatrix} + t \begin{pmatrix} 1 \\ 0 \\ 2 \end{pmatrix}$$

respectively. Show that the lines l_1 and l_2 are skew and find the acute angle between them.

11.6 Use a vector method to show that the medians of a triangle intersect at the point of trisection of the medians nearer the base.

11.7 $PQRS$ is a quadrilateral. The midpoints of PQ, QR, RS and SP are A, B, C and D respectively. Prove using vectors that $ABCD$ is a parallelogram.

11.8 A, B and C have position vectors $\mathbf{a}$, $\mathbf{b}$ and $\mathbf{c}$ respectively. P divides AB in the ratio $1 : 3$; Q divides BC in the ratio $3 : 2$. Write down the position vectors of P and Q.

If AQ and CP intersect at R, find the position vector of R.

If BR and AC intersect at S find the ratio $AS : SC$.

11.9 AB is the diameter of a circle centre O and P is any point on the circumference. Use vectors to show that $\angle APB = 90°$.

11.4 Outline Solutions to Exercises

11.1
$$|\mathbf{A} + 2\mathbf{B} - \mathbf{C}| = |9\mathbf{i} - \mathbf{j} + 5\mathbf{k}| = \sqrt{(81 + 1 + 25)} = \sqrt{107}.$$

$$\mathbf{B} \cdot (\mathbf{A} - \mathbf{C}) = (3\mathbf{i} + \mathbf{k}) \cdot (3\mathbf{i} - \mathbf{j} + 3\mathbf{k}) = 9 + 3 = 12.$$

Let $\mathbf{D} = e\mathbf{i} + f\mathbf{j} + g\mathbf{k}$ be perpendicular to $\mathbf{A}$ and $\mathbf{B}$.

Then
$$\mathbf{A} \cdot \mathbf{D} = 0 \quad \Rightarrow \quad e + 2f + 3g = 0,$$
$$\mathbf{B} \cdot \mathbf{D} = 0 \quad \Rightarrow \quad 3e \quad + g = 0.$$

To be $2\sqrt{26}$ units long, $\sqrt{(e^2 + f^2 + g^2)} = 2\sqrt{26}$.

Solve to get $\quad e = \pm 2, \quad f = \pm 8, \quad g = \mp 6, \quad$ so $\mathbf{D} = \pm(2\mathbf{i} + 8\mathbf{j} - 6\mathbf{k})$.

11.2 $\overline{AB} = \mathbf{b} - \mathbf{a} = -3\mathbf{i} + 3\mathbf{j} - 4\mathbf{k}$,
$|\overline{AB}| = \sqrt{(9 + 9 + 16)}$
$\quad = \sqrt{34}$.
Direction ratios are $-3 : 3 : -4$.

11.3 Resolved part of $\mathbf{F} = \dfrac{\mathbf{F} \cdot \mathbf{d}}{d}$.

$$\mathbf{F} \cdot \mathbf{d} = \begin{pmatrix} 2a \\ 6a \\ -6a \end{pmatrix} \cdot \begin{pmatrix} 2 \\ -1 \\ 2 \end{pmatrix} = 4a - 6a - 12a = -14a,$$

$$|\mathbf{d}| = \sqrt{(4 + 1 + 4)} = 3.$$

$$\Rightarrow \quad \text{Resolved part of } \mathbf{F} = \frac{-14a}{3} = 14 \, \text{N} \quad \Rightarrow \quad a = -3.$$

11.4 $\mathbf{p} \cdot \mathbf{q} = \mathbf{q} \cdot \mathbf{r} = \mathbf{r} \cdot \mathbf{p} = k^2/2$.

$$(\mathbf{p} + \mathbf{q} - \mathbf{r}) \cdot (\mathbf{p} - \mathbf{q} + \mathbf{r}) = p^2 - q^2 - r^2 + 2\mathbf{q} \cdot \mathbf{r}$$
$$= k^2 - k^2 - k^2 + k^2 = 0.$$

Since neither $\mathbf{p} + \mathbf{q} - \mathbf{r}$ nor $\mathbf{p} - \mathbf{q} + \mathbf{r}$ are zero vectors they must be perpendicular. Similarly for the other pairs.

$$(\mathbf{p} + \mathbf{q} - \mathbf{r}) \cdot (\mathbf{p} + \mathbf{q} - \mathbf{r}) = p^2 + q^2 + r^2 + 2\mathbf{p} \cdot \mathbf{q} - 2\mathbf{p} \cdot \mathbf{r} - 2\mathbf{q} \cdot \mathbf{r}$$
$$= 2k^2.$$

Hence $(\mathbf{p} + \mathbf{q} - \mathbf{r})^2 = 2k^2 \quad \Rightarrow \quad$ Length of $\mathbf{p} + \mathbf{q} - \mathbf{r} = k\sqrt{2}$.

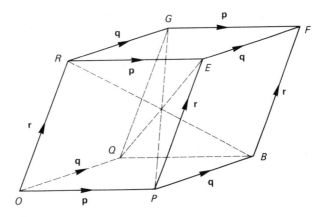

Similarly for the other two vectors.

$\mathbf{p} + \mathbf{q} - \mathbf{r}$ is the diagonal RB. Similarly, the other two vectors are the diagonals QE and PG. Hence the diagonals RB, QE and PG of the parallelopiped are mutually perpendicular and of equal length $k\sqrt{2}$.

11.5 Suppose that l_1 and l_2 intersect. Then s and t can be found so that $\mathbf{r}_1 = \mathbf{r}_2$.

$$\Rightarrow \quad 3 + 5s = -3 + t \tag{1}$$
$$1 - 2s = 5 \tag{2}$$
$$4 + 3s = 2 + 2t. \tag{3}$$

From equations 1 and 2, $s = -2$ and $t = -4$.

These values do not satisfy equation 3. Hence the lines are skew.

Directions of lines are $\begin{pmatrix} 5 \\ -2 \\ 3 \end{pmatrix}$ and $\begin{pmatrix} 1 \\ 0 \\ 2 \end{pmatrix}$.

By scalar products, $\begin{pmatrix} 5 \\ -2 \\ 3 \end{pmatrix} \cdot \begin{pmatrix} 1 \\ 0 \\ 2 \end{pmatrix} = \sqrt{(25 + 4 + 9)}\sqrt{(1 + 4)} \cos \theta$.

$$\Rightarrow \quad 11 = \sqrt{(190)} \cos \theta \quad \Rightarrow \quad \theta = 37.1°.$$

11.6 If A, B and C have position vectors $\mathbf{a}$, $\mathbf{b}$ and $\mathbf{c}$ and if D and E are the mid-points of BC and AC then

$$\overline{OE} = \mathbf{e} = \tfrac{1}{2}\mathbf{a} + \tfrac{1}{2}\mathbf{c}, \qquad \overline{OD} = \mathbf{d} = \tfrac{1}{2}\mathbf{b} + \tfrac{1}{2}\mathbf{c}.$$

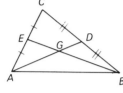

Any point on AD has position vector

$$\mathbf{r} = s\mathbf{d} + (1-s)\mathbf{a} = \frac{s}{2}(\mathbf{b}+\mathbf{c}) + (1-s)\mathbf{a}.$$

Similarly, any point on BE has position vector

$$\mathbf{r} = \frac{t}{2}(\mathbf{a}+\mathbf{c}) + (1-t)\mathbf{b}.$$

At intersection $\dfrac{t}{2} = 1 - s$ and $\dfrac{t}{2} = \dfrac{s}{2}$ $\Rightarrow$ $s = t = \tfrac{2}{3}$.

Intersection G given by $\overline{OG} = \mathbf{g} = \tfrac{1}{3}(\mathbf{a}+\mathbf{b}+\mathbf{c})$.
By symmetry this also lies on the median from C.
$\mathbf{g} = \tfrac{1}{3}\mathbf{d} + \tfrac{2}{3}\mathbf{a}$, $\Rightarrow$ G is the point of trisection of AD nearer to D.

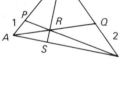

11.7 Let P, Q, R and S have position vectors $\mathbf{p}$, $\mathbf{q}$, $\mathbf{r}$ and $\mathbf{s}$ respectively.
Then $\overline{OA} = \mathbf{a} = \tfrac{1}{2}\mathbf{p} + \tfrac{1}{2}\mathbf{q}$,
$\qquad \overline{OB} = \mathbf{b} = \tfrac{1}{2}\mathbf{q} + \tfrac{1}{2}\mathbf{r}$.
$\overline{AB} = \mathbf{b} - \mathbf{a} = \tfrac{1}{2}(\mathbf{r} - \mathbf{p})$. Similarly, $\overline{DC} = \tfrac{1}{2}(\mathbf{r} - \mathbf{p})$.
Thus $\overline{AB} = \overline{DC}$, i.e. AB and DC are equal and parallel
$\Rightarrow$ $ABCD$ is a parallelogram.

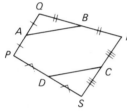

11.8 $\overline{OP} = \mathbf{p} = \tfrac{3}{4}\mathbf{a} + \tfrac{1}{4}\mathbf{b}$, $\qquad \overline{OQ} = \mathbf{q} = \tfrac{2}{5}\mathbf{b} + \tfrac{3}{5}\mathbf{c}$.
Since R lies on AQ, $\qquad \overline{OR} = \mathbf{r} = \lambda\mathbf{q} + (1-\lambda)\mathbf{a}$
$$= (1-\lambda)\mathbf{a} + \tfrac{2}{5}\lambda\mathbf{b} + \tfrac{3}{5}\lambda\mathbf{c}.$$
Similarly, since R lies on PC, $\qquad \overline{OR} = \mathbf{r} = \mu\mathbf{p} + (1-\mu)\mathbf{c}$
$$= \tfrac{3}{4}\mu\mathbf{a} + \tfrac{1}{4}\mu\mathbf{b} + (1-\mu)\mathbf{c}.$$
Equating and solving, $\lambda = \tfrac{5}{11}$ and $\mu = \tfrac{8}{11}$.
So $\overline{OR} = \mathbf{r} = \tfrac{1}{11}(6\mathbf{a} + 2\mathbf{b} + 3\mathbf{c})$.
Since S lies on BR, $\overline{OS} = \mathbf{s} = t\mathbf{r} + (1-t)\mathbf{b}$
$$= \tfrac{6}{11}t\mathbf{a} + (1 - \tfrac{9}{11}t)\mathbf{b} + \tfrac{3}{11}t\mathbf{c}.$$
But S lies on AC,
$\Rightarrow$ scalar multiple of $\mathbf{b} = 0$ $\Rightarrow$ $t = \tfrac{11}{9}$.
Hence $\mathbf{s} = \tfrac{2}{3}\mathbf{a} + \tfrac{1}{3}\mathbf{c}$, and $AS : SC = 1 : 2$.

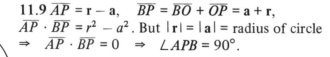

11.9 $\overline{AP} = \mathbf{r} - \mathbf{a}$, $\quad \overline{BP} = \overline{BO} + \overline{OP} = \mathbf{a} + \mathbf{r}$,
$\overline{AP} \cdot \overline{BP} = r^2 - a^2$. But $|\mathbf{r}| = |\mathbf{a}| = $ radius of circle
$\Rightarrow$ $\overline{AP} \cdot \overline{BP} = 0$ $\Rightarrow$ $\angle APB = 90°$.

12 Functions

The inverse of a one–one function. Composition of functions. Graphical illustration of a function and its inverse. Graphical solution of equations.

The approximations $\sin x \approx x$, $\tan x \approx x$, $\cos x \approx 1 - x^2/2$ (x is small and measured in radians).

12.1 Fact Sheet

(a) Notation

   (i) $f : x \to y$ is the function which maps x onto y (also $x \to y$).

  (ii) $f(x)$ is the image of x under function $f \Rightarrow f(x) = y$.

 (iii) f^{-1} is the inverse function of f.

 (iv) $f \cdot g$ is the function mapping x onto $f(x) \cdot g(x)$.

  (v) fg or $f \circ g$ is the composite function $f(g(x))$ or f operating on the result of g.

 (vi) $f \circ f$ is often written f^2.

 (vii) $f + g$ is the function mapping x onto $f(x) + g(x)$.

(viii) f' is the derived function of f.

(b) Definitions

- *Domain.* The values of x for which the function is defined. The most common of these is $\mathbb{R}$, the set of real numbers $\{x : x \in \mathbb{R}\}$.
- *Range.* The values of $f(x)$ arising from the domain.

(c) Limits of a Domain or Range

- If a function has discontinuities or is not defined for some values of x then this can be shown by inequalities.

  Example 1: $f : x \to \log x$ has a domain $\{x : x > 0\}$.

  Example 2: If $f(x) = x$, $\{x : 0 \leqslant x \leqslant 2\}$
  $$= 2x, \quad \{x : 2 < x \leqslant 4\},$$

  then the first part of the domain includes $x = 2$, but the second does not. If $x = 2$ is included in both parts the function will have two different values at that point. Since functions can only be single-valued this is not acceptable. The same function can be written

  $$f : x \to \ x \ [0, 2]$$
  $$f : x \to 2x \ (2, 4].$$

[ ] imply that the limit points are included.
( ) imply that the limit points are not included.

Graphically, included limit points are represented by ●, excluded limit points by ○.

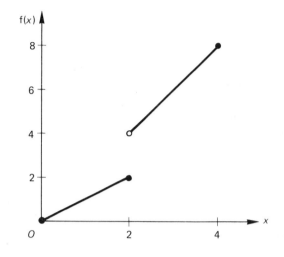

- A function and its inverse can be recognized graphically since the reflection of $f(x)$ in the line $y = x$ gives $f^{-1}(x)$.
- If the inverse is to be a function its domain may have to be restricted. For example the inverse function of $\cos x$ is $\cos^{-1} x$ with domain $\{x : -1 \leqslant x \leqslant 1\}$.

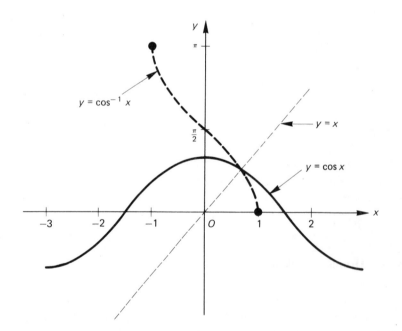

(d) Graphical Solutions of Equations

If an equation $y = f(x)$ is complicated, split $f(x)$ into $g(x) + h(x)$ where g and h are easily sketched. Then the solutions of $f(x) = 0$ are given by the points of intersection of the graphs $y = g(x)$ and $y = -h(x)$.

For example, if $y = 2 \sin x - x$ then $y = 0$ when $y = 2 \sin x$ and $y = +x$ intersect.

130

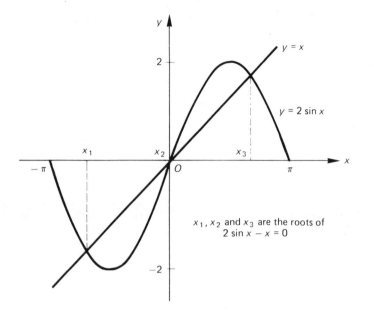

x_1, x_2 and x_3 are the roots of
$2 \sin x - x = 0$

(e) Odd and Even Functions

- *Definition:*
  If $f(-x) = f(x)$ then $f(x)$ is an even function.
  If $f(-x) = -f(x)$ then $f(x)$ is an odd function.
- *Properties:*
  Even functions are symmetrical about the line $x = 0$.
  Odd functions have half-turn rotational symmetry about O.
  A function containing only even powers of x (including x^0) is an even function, but one containing only odd functions is not necessarily odd.
- *Examples:*

Even: Odd:

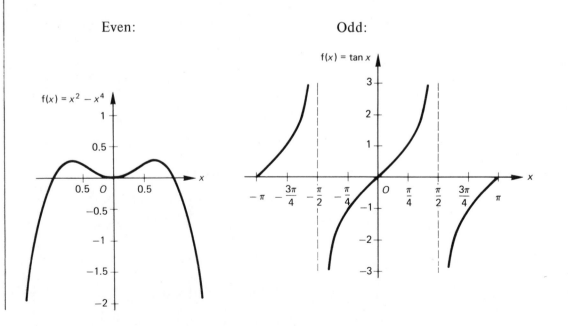

Neither even nor odd:

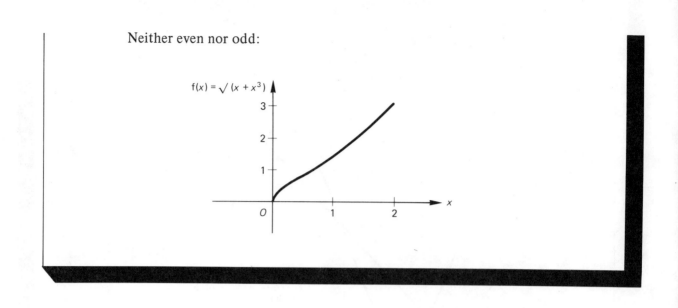

12.2 Worked Examples

12.1 If $f : x \to 2x - 3$ and $g : x \to 3x^2 + 2x$ are functions from $\mathbb{R}$ to $\mathbb{R}$, then $(g' \circ f)$ maps x onto

A, $12x + 1$;   B, $12x + 4$;   C, $12x - 16$;
D, 16;   E, none of these.

● $g' : x \to 6x + 2$,
   $g' \circ f : x \to 6(2x - 3) + 2 = 12x - 16$.   Answer **C**

12.2 The functions f and g are defined by $f(x) = x^2$, $g(x) = x + 1$ and each have domain the positive real numbers $\mathbb{R}^+$. Express the following in terms of f and g:

(i) $x - 2$;   (ii) $x + 2$;
(iii) $x^2 + 1$;   (iv) $(x + 1)^2$;
(v) $(x - 2)^4$.

● $f(x) = x^2$, $g(x) = x + 1$, $g^{-1}(x) = x - 1$.

(i) $x - 2 = g^{-1}(g^{-1}(x))$ or g^{-2};   (ii) $x + 2 = g(g(x))$ or g^2;
(iii) $x^2 + 1 = g(f(x))$ or gf;   (iv) $(x + 1)^2 = f(g(x))$ or fg;
(v) $(x - 2)^4 = ff\, g^{-1}g^{-1}\,(x)$ or $f^2 g^{-2}$.

12.3 The functions $g(x)$ and $h(x)$ are defined by

$$g(x) = \sqrt{(9 - x^2)}\quad |x| \leqslant 3; \quad h(x) = 2 + x^2.$$

If $g \circ h$ exists give the domain and range of $h(x)$ and $g \circ h$.

● $g \circ h = \sqrt{[9 - (2 + x^2)^2]}$ requires that $9 - (2 + x^2)^2 \geqslant 0$
   $\Rightarrow\ 9 \geqslant (2 + x^2)^2$ so $x^2 \leqslant 1$,   i.e. $|x| \leqslant 1$.
   For $g \circ h$ to exist, the domain of $h(x)$ is $|x| \leqslant 1$, and the range of $h(x)$ is $2 \leqslant h(x) \leqslant 3$.
   The domain of $g \circ h$ is $|x| \leqslant 1$, the range of $g \circ h$ is $0 \leqslant g \circ h \leqslant \sqrt{5}$.

132

12.4 A function with domain and range [0, 5] is defined by

$$f(x) = x + 3 \qquad \text{if } x \in [0, 2],$$
$$= 5 - x \qquad \text{if } x \in (2, 5].$$

Sketch the graphs of f, f^{-1} and f $\circ$ f and state clearly the ranges of each function and the points on the graphs of any discontinuity.

$$f(x) = x + 3 \qquad \text{if } x \in [0, 2],$$
$$= 5 - x \qquad \text{if } x \in (2, 5].$$

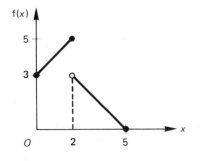

Domain [0, 2] gives range [3, 5].
Domain (2, 5] gives range (3, 0].
Range of $f(x)$ is [0, 5] with discontinuity at $x = 2$.

$$f^{-1}(x) = 5 - x \qquad \text{if } x \in [0, 3),$$
$$= x - 3 \qquad \text{if } x \in [3, 5].$$

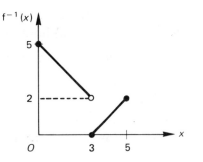

Range of $f^{-1}(x)$ is [0, 5] with discontinuity at $x = 3$.

Approach f $\circ$ f with care, especially at discontinuities.

Range of f becomes the domain for the second function.

Domain [0, 2] → range [3, 5]; domain [3, 5] → range [2, 0].
Domain (2, 3) → range (3, 2); domain (3, 2) → range (2, 3).
Domain [3, 5] → range [2, 0]; domain [2, 0] → range [5, 3].

$$f \circ f = 5 - (x + 3) = 2 - x \qquad \text{if } x \in [0, 2]$$
$$= 5 - (5 - x) = x \qquad \text{if } x \in (2, 3)$$
$$= (5 - x) + 3 = 8 - x \qquad \text{if } x \in [3, 5].$$

Range of f $\circ$ f is [0, 5], with discontinuities at (2, 0) and (3, 5).

133

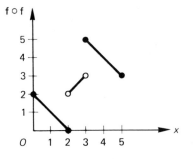

12.5 State the largest possible real domains and ranges of the functions f and g defined by

$$f(x) = \sqrt{(4 - 9x^2)}, \qquad g(x) = \cos x.$$

Find the composite function gf and state its domain and range.
Can fg be formed? Give reasons for your answer.

● $f(x) = \sqrt{(4 - 9x^2)}$ has domain $\{x : -\frac{2}{3} \leqslant x \leqslant \frac{2}{3}\}$,
range $0 \leqslant f(x) \leqslant 2$.
$g(x) = \cos x$ has domain $\{x : x \in \mathbb{R}\}$, range $-1 \leqslant g(x) \leqslant 1$.
$gf = \cos(\sqrt{(4 - 9x^2)})$ has domain $\{x : -\frac{2}{3} \leqslant x \leqslant \frac{2}{3}\}$,
range $\cos 2 \leqslant gf \leqslant \cos 0 = 1$.
$fg = \sqrt{(4 - 9\cos^2 x)}$, if it exists. However it is not defined for $|\cos x| > \frac{2}{3} \Rightarrow$ it is not defined for the full range of $g(x)$.
Therefore fg can only be formed for domain $\{x : |\cos x| \leqslant \frac{2}{3}\}$.

12.6 For each of the following functions state, without proof, if it is: (i) even or odd or neither; and (ii) bounded or not bounded:

(a) $f_1(x) = \cos x - \cos^3 x$, (b) $f_2(x) = \dfrac{x^2}{x^2 + 2}$, (c) $f_3(x) = x \cos x$,

(d) $f_4(x) = \sqrt{(x + x^4)}$.

● (a) $\cos x$ and $\cos^3 x$ are both even functions and bounded.
Therefore $f_1(x)$ is even and bounded.

(b) x^2 and $x^2 + 2$ are both even, therefore the quotient is even. x^2 and $x^2 + 2$ are both unbounded but the quotient is bounded (see sketch). Therefore $f_2(x)$ is even and bounded.

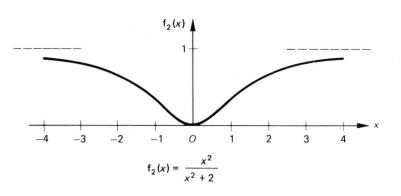

$$f_2(x) = \frac{x^2}{x^2 + 2}$$

134

(c) x is odd, cos x is even, therefore the product is odd. cos x is bounded but x is unbounded, therefore the product is unbounded.
Therefore $f_3(x)$ is odd and unbounded.

(d) This square root function is neither odd nor even. $x + x^4$ is unbounded, so $f_4(x)$ is unbounded.
Therefore $f_4(x)$ is neither even nor odd and is unbounded.

12.7 $f(x) = -\sqrt{(9 - 25x^2)}$ $(0 \leqslant x \leqslant \frac{3}{5})$, where $\sqrt{\ }$ denotes the positive square root. Find an expression for $f^{-1}(x)$ and write down the domain and range for $f^{-1}(x)$. Sketch $f(x)$ and $f^{-1}(x)$ on the same axes.

- **Draw a flowchart for $f(x)$ if in doubt.**

$$x \boxed{\text{square}} \rightarrow x^2 ; \boxed{\text{multiply by 25}} \rightarrow 25x^2 ;$$

$$\boxed{\text{subtract from 9}} \rightarrow 9 - 25x^2 ; \boxed{\text{square root}} \rightarrow \sqrt{(9 - 25x^2)};$$

$$\boxed{\text{multiply by } -1} \rightarrow -\sqrt{(9 - 25x^2)}.$$

For the inverse,

$$x \boxed{\text{divide by } -1} \rightarrow -x; \boxed{\text{square}} \rightarrow x^2 ;$$

$$\boxed{\text{subtract from 9}} \rightarrow 9 - x^2 ; \boxed{\text{divide by 25}} \rightarrow \left(\frac{9 - x^2}{25}\right) ;$$

$$\boxed{\text{square root}} \rightarrow \sqrt{\left(\frac{9 - x^2}{25}\right)}.$$

$f^{-1}(x) = \frac{1}{5}\sqrt{(9 - x^2)}$. Domain $\{-3 \leqslant x \leqslant 0\}$, range $\{0 \leqslant f^{-1}(x) \leqslant \frac{3}{5}\}$.

Notice the reflection in line $y = x$.

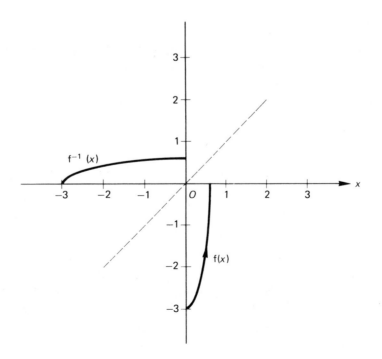

135

12.8 Prove that when θ is sufficiently small,

$$\theta \approx \sin \theta \approx \tan \theta.$$

When viewed from a stationary boat lying in the path of a ship of breadth 20 m the port and starboard lights subtend an angle of $36'$. Find the distance of the lights from the observation point. After 2 min the angle has increased to $1° \; 12'$. Calculate the speed of the ship in km h^{-1}.

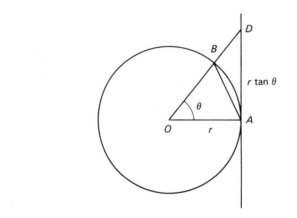

From the figure: area of triangle $OAB = \frac{1}{2}r^2 \sin \theta$.   (A1)
Area of sector $OAB = \frac{1}{2}r^2 \theta$.   (A2)
Area of triangle $OAD = \frac{1}{2}r^2 \tan \theta$.   (A3)
When θ is acute, A1 $\leqslant$ A2 $\leqslant$ A3
$\Rightarrow \quad \sin \theta \leqslant \theta \leqslant \tan \theta.$
As $\theta \to 0$, A1 $\to$ A2 $\to$ A3,
i.e. for small θ, $\theta \approx \sin \theta \approx \tan \theta$.

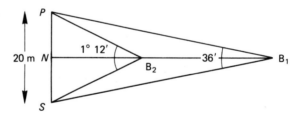

Let N be the midpoint of PS, B_1 and B_2 the two positions of the boat relative to the ship.

$$\frac{PN}{NB_1} = \tan 18' \approx \left(\frac{18}{60}\right)\left(\frac{\pi}{180}\right)$$

so $10 = \dfrac{\pi}{600} NB_1,\quad \Rightarrow\quad NB_1 = \dfrac{6000}{\pi}.$

Two minutes later $\dfrac{PN}{NB_2} = \tan 36' \approx \left(\dfrac{36}{60}\right)\left(\dfrac{\pi}{180}\right),$

so $10 = \dfrac{\pi}{300} NB_2\quad \Rightarrow\quad NB_2 = \dfrac{3000}{\pi}.$

Distance travelled by ship in two minutes is

$$NB_1 - NB_2 = \frac{6000 - 3000}{\pi} \text{ m}, = 955 \text{ m} = 0.955 \text{ km}.$$

Speed is $30 \times 0.955 = 28.6$ km h^{-1}.

12.3 Exercises

12.1 The functions f and h each have domain the positive real numbers and are defined by

$$f(x) = 2/x, \qquad h(x) = x + 1.$$

Express in terms of f and h the functions (a) $\dfrac{2}{x + 2}$, (b) $\dfrac{3 + x}{1 + x}$.

Give the domain and range of each of the functions.

12.2 The functions f and g are defined by $f(x) = \cos x$, $g(x) = \sqrt{(1 - 9x^2)}$. Find the composite function fg and state its domain and range. Can gf be formed? Give reasons for your answer.

12.3 Which of the following drawings is most likely to show the graph of the function $f : x \rightarrow \log (x^2 + 1)$?

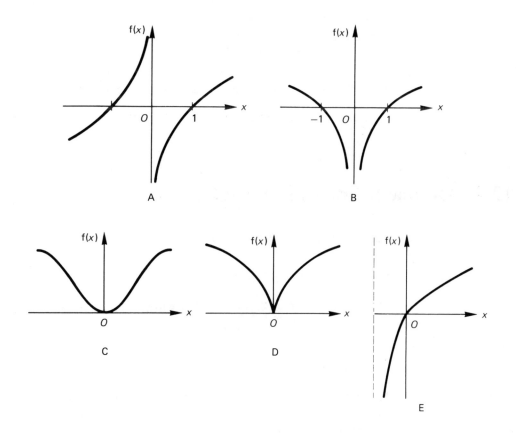

12.4 A function with domain [0, 4] is defined by

$$f(x) = x + 3 \qquad \text{if } x \in [0, 1]$$

$$= \frac{4}{x} - 1 \qquad \text{if } x \in (1, 4].$$

Sketch the graphs of f, f^{-1}, f $\circ$ f, showing clearly the point of the graphs at any discontinuity.

12.5 Sketch the graphs of the functions and their inverses on the same axes:
(a) $f(x) = x^2 - 2$ $\{0 \leqslant x \leqslant 3\}$;
(b) $f(x) = e^{1-x}$ $\{0 \leqslant x \leqslant 3\}$;
(c) $f(x) = \sin(x/2)$ $\{0 \leqslant x \leqslant 1\}$.

12.6 Which of the following statements is true for the functions defined by

$$f(x) = x - \frac{1}{x}, \quad x > 1$$

$$g(x) = \frac{2}{1-x}, \quad x > 1?$$

A, f is increasing, g is decreasing;
B, f is increasing, g is increasing;
C, f is decreasing, g is decreasing;
D, f is decreasing, g is increasing;
E, none of these.

12.7 The sun is approximately 150 million km from an earthbound observer who estimates that its diameter subtends an angle of 0.5°. Estimate its diameter and hence the volume of the sun.

12.8 Plot on the same axes the graphs $y = 2 + x^2$ and $y = \dfrac{1}{x}$.

Hence estimate the value of the real root of the equation $x^3 + 2x - 1 = 0$.

12.4   Outline Solutions to Exercises

12.1 (a) $f(x) = \dfrac{2}{x}$, $h(x) = x + 1$, $h^2 = x + 2$ $\Rightarrow$ $\dfrac{2}{x+2} = fh^2$;

domain $(0, \infty)$, range $(0, 1)$.

(b) $\dfrac{3+x}{1+x} = 1 + \dfrac{2}{x+1} = 1 + (fh) = hfh$; domain $(0, \infty)$, range $(1, 3)$.

12.2 $f(x) = \cos x$, $g(x) = \sqrt{(1 - 9x^2)}$ $\Rightarrow$ $fg = \cos[\sqrt{(1 - 9x^2)}]$.
Domain of $f(x)$ is $x \in \mathbb{R}$, domain of $g(x)$ is $-\frac{1}{3} \leqslant x \leqslant \frac{1}{3}$,
$\Rightarrow$ domain of $fg(x)$ is $-\frac{1}{3} \leqslant x \leqslant \frac{1}{3}$.
Range of $fg(x)$ is $\cos(1) \leqslant fg \leqslant 1$.
$gf(x) = \sqrt{(1 - 9 \cos^2 x)}$ defined for $|\cos x| \leqslant \frac{1}{3}$ thus $gf(x)$ exists only if the domain of $f(x)$ is restricted to

$$\cos^{-1}\left(\tfrac{1}{3}\right) \leqslant x \leqslant \cos^{-1}\left(-\tfrac{1}{3}\right).$$

12.3 $f : x \rightarrow \log(x^2 + 1)$ is defined for all values of x.
This excludes answers A, B and E.

$$f'(x) = \frac{2x}{(x^2 + 1) \ln 10}, \text{ when } x = 0, f'(x) = 0.$$

Graph C

12.4

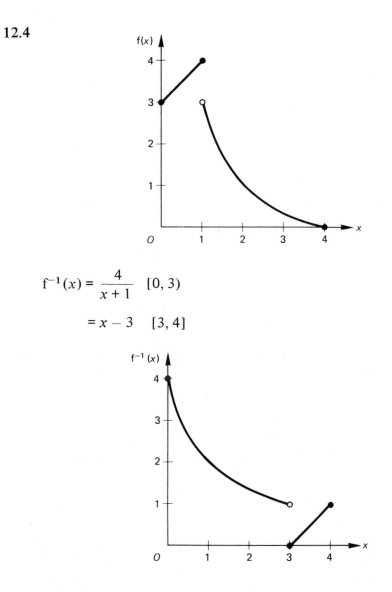

$$f^{-1}(x) = \frac{4}{x + 1} \quad [0, 3)$$

$$= x - 3 \quad [3, 4]$$

$f(x) = 1$ when $x = 2$. The domain of $f \circ f$ will have discontinuities at $x = 1$ and $x = 2$.

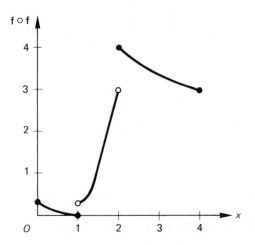

$$f \circ f = \frac{4}{x+3} - 1 = \frac{1-x}{x+3} \qquad [0, 1]$$

$$= \frac{4}{\frac{4}{x} - 1} - 1 = \frac{5x-4}{4-x} \qquad (1, 2)$$

$$= \frac{4}{x} - 1 + 3 = \frac{4}{x} + 2 \qquad [2, 4]$$

12.5

(a) $f(x) = x^2 - 2$, $f^{-1}(x) = \sqrt{(x+2)}$;

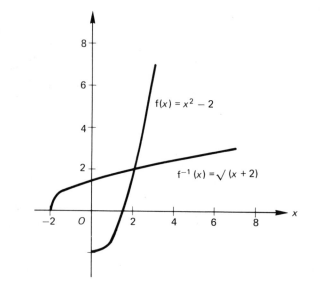

(b) $f(x) = e^{1-x}$, $f^{-1}(x) = 1 - \ln x$;   (c) $f(x) = \sin\left(\frac{x}{2}\right)$, $f^{-1}(x) = 2 \sin^{-1} x$.

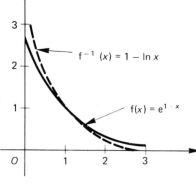 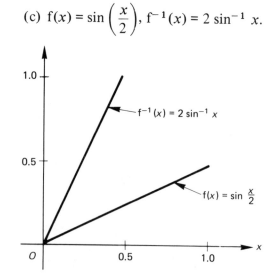

12.6 $f(x) = x - \frac{1}{x}$, $x > 1$; $f'(x) = 1 + \frac{1}{x^2}$,

$g(x) = \frac{2}{1-x}$, $x > 1$; $g'(x) = \frac{2}{(1-x)^2}$

$f'(x)$ and $g'(x)$ are both positive for $x > 1$.

                                                        <u>Answer **B**</u>

12.7 Angle $TOS = 0.25°$, $ST = 150 \times 10^6 \times \tan(0.25°) \approx 0.655 \times 10^6$ km.
Diameter is approx. 1.31×10^6 km.
Volume $= \frac{4}{3}\pi r^3 = 1.2 \times 10^{18}$ km^3.

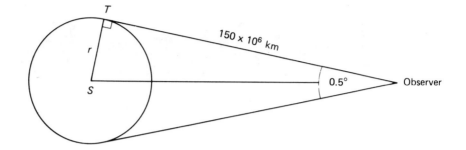

12.8 From the graph, $x = 0.45$ or 0.46.

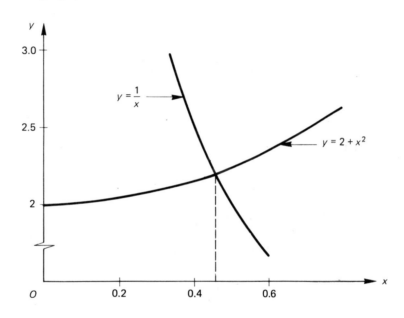

13 Exponential and Logarithmic Functions

The exponential and logarithmic functions and their simple properties. $a^x = e^{x \ln a}$.

13.1 Fact Sheet

a, b and c are positive throughout this fact sheet.

(a) Exponential Functions

- *Definition:* a^x is an exponential function.
- *Properties:*
  - (i) $a^x > 0$ for all $x \in \mathbb{R}$;
  - (ii) $a^x = 1$ when $x = 0$;
  - (iii) $a^x \to 0$ as $x \to -\infty$ $(a > 1)$;
  - (iv) $a^x \to \infty$ as $x \to \infty$ $(a > 1)$.

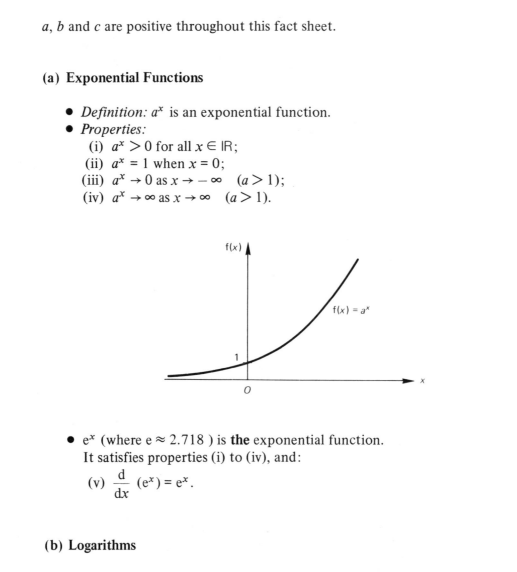

- e^x (where $e \approx 2.718$) is **the** exponential function. It satisfies properties (i) to (iv), and:
  - (v) $\dfrac{d}{dx}(e^x) = e^x$.

(b) Logarithms

- *Definition:* If $b = a^c$, then $\log_a b = c$. $(a \neq 1)$

- *Properties:*
  - (i) $\log_a bc = \log_a b + \log_a c$;
  - (ii) $\log_a \dfrac{b}{c} = \log_a b - \log_a c$;
  - (iii) $\log_a (b^c) = c \log_a b$;
  - (iv) $\log_a b = \dfrac{\log_c b}{\log_c a}$;
  - (v) $\log_a b = \dfrac{1}{\log_b a}$.
- Common logarithms are logarithms to base 10.
  $\log_{10} a$ is sometimes written $\log a$, or $\lg a$.

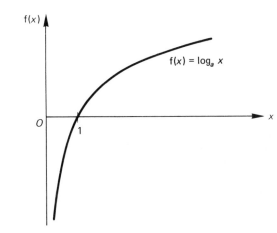

- Natural or Naperian logarithms are logarithms to base e, written $\log_e a$ or ln a.
  $a^x = e^{x \ln a}$.

(c) **Series**

$$e^x = 1 + x + \frac{x^2}{2!} + \frac{x^3}{3!} + \frac{x^4}{4!} + \ldots + \frac{x^n}{n!} + \ldots$$

$$e^{-x} = 1 - x + \frac{x^2}{2!} - \frac{x^3}{3!} + \frac{x^4}{4!} - \ldots + (-1)^n \frac{x^n}{n!} \ldots +$$

$$\ln (1 + x) = x - \frac{x^2}{2} + \frac{x^3}{3} - \frac{x^4}{4} + \ldots + (-1)^{n-1} \frac{x^n}{n} + \ldots \quad (-1 < x \leqslant 1)$$

$$\ln (1 - x) = -x - \frac{x^2}{2} - \frac{x^3}{3} - \frac{x^4}{4} - \ldots - \frac{x^n}{n} - \ldots \quad (-1 \leqslant x < 1).$$

13.2 Worked Examples

13.1 The graph of $y = \log_3 x$ is the image of the graph of $y = 3^x$ under a reflection in

A, the x-axis;   B, the y-axis;   C, the line $y = x$;
D, the line $y = -x$;   E, none of these.

- 3^x is the inverse function of $\log_3 x$.
  Hence reflection is in the line $y = x$.   Answer **C**

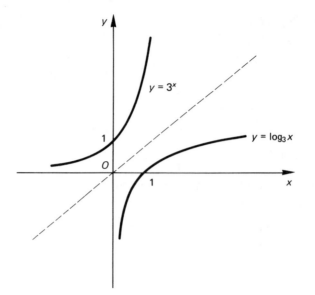

13.2 Given that $\log 2 = a$, $\log 3 = b$, $\log 5 = c$, then $\log 750$ can be expressed in terms of a, b and c as

A, $a + b + c$;   B, $a + 3b + c$;   C, $2a + 2b + 2c$;
D, $a + b + 3c$;   E, none of these.

● $750 = (5)\,(5)\,(5)\,(3)\,(2) = 5^3\,(3)\,(2)$,
so $\log 750 = 3\log 5 + \log 3 + \log 2 = 3c + b + a$.   **Answer D**

13.3 Prove that $\log_b a = \dfrac{1}{\log_a b}$.

By using the substitution $y = \log_x 4$, or otherwise, solve the equation
$4\log_{16} x - 1 = \log_x 4$.

● If $c = \log_b a$ then $a = b^c$.
Taking logarithms to base a, $\log_a a = c\log_a b$.

But $\log_a a = 1$ $\Rightarrow$ $1 = c\log_a b$ or $c = \dfrac{1}{\log_a b}$.

Hence $\log_b a = \dfrac{1}{\log_a b}$.

$4\log_{16} x - 1 = \log_x 4$.

$\log_{16} x = \dfrac{1}{\log_x 16} = \dfrac{1}{(2\log_x 4)}$.

Substituting in the equation, $\dfrac{4}{2\log_x 4} - 1 = \log_x 4$.

Let $\log_x 4 = y$, then $\dfrac{4}{2y} - 1 = y$.

Multiplying by y, $y^2 + y - 2 = 0$,

$(y + 2)\,(y - 1) = 0$ $\Rightarrow$ $y = 1$ or -2.

If $\log_x 4 = 1$ then $x = 4$.
If $\log_x 4 = -2$ then $4 = x^{-2}$ $\Rightarrow$ $x = \pm 0.5$.
Log x is only defined for positive values of x.
Therefore $x = 4$ or 0.5.

13.4 Write down the first 4 terms in the expansion of e^x. Obtain the first 4 terms in the expansion of $(x^2 + x)e^x$. Deduce that $(x^2 + x)(e^x - 1)$ is always positive when x is positive.

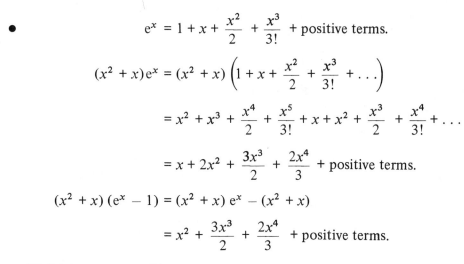

●
$$e^x = 1 + x + \frac{x^2}{2} + \frac{x^3}{3!} + \text{positive terms.}$$

$$(x^2 + x)e^x = (x^2 + x)\left(1 + x + \frac{x^2}{2} + \frac{x^3}{3!} + \ldots\right)$$

$$= x^2 + x^3 + \frac{x^4}{2} + \frac{x^5}{3!} + x + x^2 + \frac{x^3}{2} + \frac{x^4}{3!} + \ldots$$

$$= x + 2x^2 + \frac{3x^3}{2} + \frac{2x^4}{3} + \text{positive terms.}$$

$$(x^2 + x)(e^x - 1) = (x^2 + x)e^x - (x^2 + x)$$

$$= x^2 + \frac{3x^3}{2} + \frac{2x^4}{3} + \text{positive terms.}$$

All the terms are positive.
Therefore $(x^2 + x)(e^x - 1)$ is always positive, provided x is positive.

13.5 The graph illustrates the law $y = ax^b$. The value of b is nearest to
A, -0.5;   B, 2;   C, 1.2;   D, 4;   E, 16.

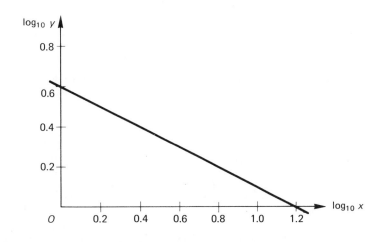

● $y = ax^b$. Take logarithms to base 10.
$\log_{10} y = \log_{10} ax^b = \log_{10} a + b \log_{10} x$
or $Y = A + bX$,
where $Y = \log_{10} y$, $A = \log_{10} a$, b is the gradient of the line, and $X = \log_{10} x$.
From the diagram, the gradient of the line is $\dfrac{-0.6}{1.2} = -0.5$.

<u>Answer A</u>

13.6 Find the first three non-zero terms in the expansion of $\ln[(1 + 3x)(1 - 2x)]$ in a series of ascending powers of x.

● $\ln[(1 + 3x)(1 - 2x)] = \ln(1 + 3x) + \ln(1 - 2x)$.
$\ln(1 + 3x) = (3x) - \dfrac{(3x)^2}{2} + \dfrac{(3x)^3}{3} - \ldots$

$$= 3x - \frac{9x^2}{2} + \frac{27x^3}{3} - \ldots \qquad (1)$$

$$\ln(1 - 2x) = (-2x) - \frac{(-2x)^2}{2} + \frac{(-2x)^3}{3} - \ldots$$

$$= -2x - \frac{4x^2}{2} - \frac{8x^3}{3} - \ldots \qquad (2)$$

Adding expansions 1 and 2,

$$\ln[(1 + 3x)(1 - 2x)] = x - \frac{13x^2}{2} + \frac{19x^3}{3} + \ldots$$

13.7 Find the first two non-zero terms in the expansion in ascending powers of x, of $e^{2x} - e^{-2x}$. Given that x is so small that its fourth and higher powers may be neglected, find the numerical values of the constants a, b and c in the approximate formula

$$\log_e(e^{2x} - e^{-2x}) \approx \log_e x + a + bx + cx^2 \qquad (x > 0).$$

●
$$e^{2x} = 1 + (2x) + \frac{(2x)^2}{2!} + \frac{(2x)^3}{3!} + \frac{(2x)^4}{4!} + \ldots$$

$$e^{-2x} = 1 + (-2x) + \frac{(-2x)^2}{2!} + \frac{(-2x)^3}{3!} + \frac{(-2x)^4}{4!} + \ldots$$

Subtracting $\quad e^{2x} - e^{-2x} = 2\left(2x + \frac{8x^3}{6} + \ldots\right)$

$$= 4x + \frac{8x^3}{3} + \ldots$$

$$\log_e(e^{2x} - e^{-2x}) \approx \log_e\left(4x + \frac{8x^3}{3}\right) = \log_e\left[4x\left(1 + \frac{2x^2}{3}\right)\right]$$

$$= \log_e 4x + \log_e\left(1 + \frac{2x^2}{3}\right).$$

But $\log_e 4x = \log_e x + \log_e 4$,

and $\log_e\left(1 + \frac{2x^2}{3}\right) = \frac{2x^2}{3} + $ higher powers of x^2.

Thus $\qquad \log_e(e^{2x} - e^{-2x}) = \log_e x + \log_e 4 + \frac{2x^2}{3} + \ldots$

$a = \log_e 4$, $b = 0$, $c = \frac{2}{3}$.

13.8 When x is so small that x^3 and higher powers of x can be ignored, find the values of a and b such that $\ln\left(\dfrac{1 + ax}{1 + bx}\right) = e^x - e^{2x}$.

● $\ln\left(\dfrac{1 + ax}{1 + bx}\right) = \ln(1 + ax) - \ln(1 + bx)$

$$= \left(ax - \frac{a^2 x^2}{2}\right) - \left(bx - \frac{b^2 x^2}{2}\right) + \text{higher powers of } x$$

$$= x(a - b) + \frac{x^2}{2}(b^2 - a^2) + \ldots$$

$$e^x - e^{2x} = \left[1 + x + \frac{x^2}{2!}\right] - \left[1 + (2x) + \frac{(2x)^2}{2!}\right]$$

+ higher powers of x.

$$= -x - \frac{3x^2}{2} - \ldots.$$

If $\ln\left(\dfrac{1 + ax}{1 + bx}\right) = e^x - e^{2x}$,

then
$$a - b = -1 \qquad\qquad\qquad (1)$$

and
$$b^2 - a^2 = -3. \qquad\qquad\qquad (2)$$

Factorizing equation 2 and using equation 1 gives $a + b = -3$.
Adding to equation 1 gives $a = -2$, $b = -1$.

13.9 Draw the graph of $y = \log_e\left(\dfrac{3 + x}{3 - x}\right)$ for $-2.5 \leqslant x \leqslant 2.5$.

Use your graph to estimate the root of the equation
$$4 - \frac{3x}{2} = \log_e\left(\frac{3 + x}{3 - x}\right) \text{ to one decimal place.}$$

●

x	-2.5	-1.5	-0.5	0.5	1.5	2.5
$\dfrac{3 + x}{3 - x}$	0.091	0.333	0.714	1.4	3	11
$\log_e\left(\dfrac{3 + x}{3 - x}\right)$	-2.40	-1.10	-0.34	0.34	1.10	2.40

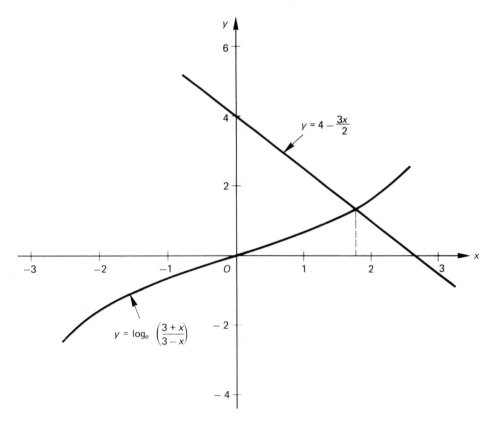

From the graph, $x = 1.8$.

13.3 Exercises

13.1 Find the values of a and b, such that

$$a = 3b \text{ and } \log_3 a + \log_3 b = 2.$$

13.2 Given that $\ln 2 = a$, $\ln 3 = b$ and $\ln 5 = c$, express in terms of a, b and c:

(a) $\ln 7.5$;   (b) $\ln 125$;   (c) $\log_2 e^2$;   (d) $\dfrac{\log_3 2}{\log_5 3}$;   (e) $\sqrt{30}$.

13.3 Solve, for real x, the equation $\log_3 (x + 8) = 2 - \log_3 x$.

13.4 Given that $2 (\lg y) + 3 (\lg x) = 4$, express y in terms of x in a form not involving logarithms ($\lg \equiv \log_{10}$).

13.5 A relation of the form $y = ab^x - 3$ is known to exist between two variables x and y. By drawing a suitable linear graph, or otherwise, use the following table of approximate values of x and y to estimate the values of a and b:

x	2	3	3.5	4	4.5	5
y	3	11	18	30	48	76

13.6 For each of the functions f, g, and h defined below determine whether it is (a) odd, (b) even, (c) neither. Give reasons for your answers.

$$f(x) = \frac{2e^x}{1 + e^{2x}}; \qquad g(x) = \ln \frac{1 + x}{1 - x}; \qquad h(x) = \cos 3x + \sin 3x.$$

13.7 Expand $\dfrac{e^{2x+1}}{1 - 3x}$ as a series of ascending powers of x as far as the term in x^3 and give the set of values of x for which the expansion is valid.

13.8 Obtain an approximate value for the positive root of the equation $(x + 1) (2 - x) = 5 \ln (1 + x)$, (a) by drawing the graphs of $y = (x + 1) (2 - x)$ and $y = 5 \ln (1 + x)$ for values of x between $x = 0$ and $x = 2$, and (b) by expressing $\ln (1 + x)$ as a series of ascending powers of x as far as the term in x^2.

13.4 Outline Solutions to Exercises

13.1 $\log_3 a + \log_3 b = 2 \Rightarrow \log_3 ab = 2 \Rightarrow ab = 3^2 = 9$.
But $a = 3b$, so $a^2 = 27 \Rightarrow a = \pm 3\sqrt{3}, b = \pm\sqrt{3}$.
But only logs of positive numbers are defined,
$a = 3\sqrt{3}, b = \sqrt{3}$.

13.2

(a) $\ln (7.5) = \ln \dfrac{(5)(3)}{2} = \ln 5 + \ln 3 - \ln 2 = c + b - a.$

(b) $\ln (125) = \ln (5^3) = 3 \ln 5 = 3c.$

(c) $\log_2 e^2 = 2 \log_2 e = \dfrac{2}{\ln 2} = \dfrac{2}{a}.$

(d) $\log_3 2 = \dfrac{\ln 2}{\ln 3} = \dfrac{a}{b};\ \log_5 3 = \dfrac{\ln 3}{\ln 5} = \dfrac{b}{c}.$

$\Rightarrow\ \dfrac{\log_3 2}{\log_5 3} = \dfrac{a/b}{b/c} = \dfrac{ac}{b^2}.$

(e) $\sqrt{30} = \sqrt{(2 \times 3 \times 5)}.$
$\ln \sqrt{30} = \tfrac{1}{2} \ln 30 = \tfrac{1}{2}(a + b + c) \quad \Rightarrow \quad \sqrt{30} = e^{(a + b + c)/2}.$

13.3 $\log_3 (x + 8) + \log_3 x = 2 \quad \Rightarrow \quad \log_3 x (x + 8) = 2$
$\Rightarrow\ x (x + 8) = 3^2 = 9 \quad \Rightarrow \quad x = -9 \text{ or } x = 1.$
But x must be positive, so $x = 1$.

13.4 $\lg y + 3 \lg x = 4 = \lg (10^4).$
Then $\lg (y^2 x^3) = \lg (10^4) \quad \Rightarrow \quad y^2 x^3 = 10^4 \quad \Rightarrow \quad y^2 = 10^4 x^{-3}.$
y must be positive, so $y = 100 x^{-3/2}.$

13.5 $y + 3 = ab^x \quad \Rightarrow \quad \log (y + 3) = x \log b + \log a.$
Plotting $Y = \log (y + 3)$ against x gives a straight line.
From the graph, gradient $= 0.38 = \log b, \quad \Rightarrow \quad b = 2.4.$
Y-intercept $= 0 = \log a \quad \Rightarrow \quad a = 1.$

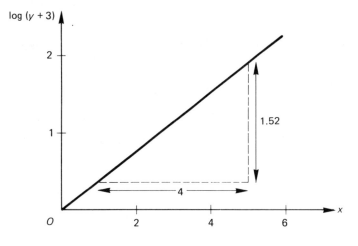

13.6

$f(x) = \dfrac{2}{e^{-x} + e^x}$, $f(-x) = \dfrac{2}{e^x + e^{-x}} = f(x).$ Therefore $f(x)$ is even.

$g(x) = \ln (1 + x) - \ln (1 - x),\ g(-x) = \ln (1 - x) - \ln (1 + x) = -g(x).$
Therefore $g(x)$ is odd.

$h(x) = \cos 3x + \sin 3x,\ h(-x) = \cos 3x - \sin 3x.$
Therefore $h(x)$ is neither even nor odd.

13.7 $f(x) = \dfrac{e^{2x+1}}{1 - 3x}$.

$e^{2x+1} = e(e^{2x}) = e\left[1 + (2x) + \dfrac{(2x)^2}{2!} + \dfrac{(2x)^3}{3!} + \ldots\right]$

$= e\left(1 + 2x + 2x^2 + \dfrac{4x^3}{3} + \ldots\right)$ \qquad (valid for all x).

$(1 - 3x)^{-1} = 1 + 3x + 9x^2 + 27x^3 + \ldots$ (valid for $|x| < \frac{1}{3}$).

$f(x) = e\left(1 + 5x + 17x^2 + \dfrac{157}{3}x^3 + \ldots\right)$ (valid for $|x| < \frac{1}{3}$).

13.8

(a) From the graph, $x \approx 0.6$.

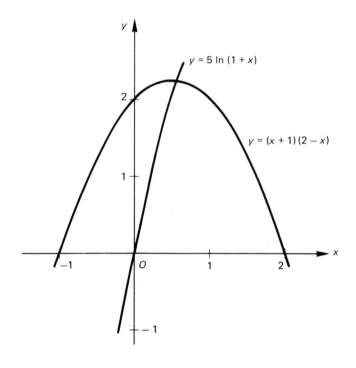

(b) $(x + 1)(2 - x) = 5\left(x - \dfrac{x^2}{2} + \text{higher powers of } x\right)$ $(-1 < x \leqslant 1)$.

Neglecting the higher powers of x: $2 + x - x^2 = 5x - \dfrac{5x^2}{2}$.

$$3x^2 - 8x + 4 = 0 \quad \Rightarrow \quad x = \tfrac{2}{3} \text{ or } 2.$$

But the log series is not valid for $x = 2$. Therefore $x = \tfrac{2}{3}$.

14 Differentiation—1

The idea of a limit and a derivative defined as a limit. The gradient of a tangent as the limit of the gradient of a chord.
Differentiation of standard functions. Differentiation of sum, product and quotient of functions, and of a composition of functions.

14.1 Fact Sheet

If $y = f(x)$ then $\dfrac{dy}{dx} = f'(x) = \lim\limits_{\delta x \to 0} \dfrac{f(x + \delta x) - f(x)}{\delta x}$

$$\text{or } f'(a) = \lim\limits_{b \to a} \dfrac{f(a) - f(b)}{a - b}.$$

(a) Product

$$\frac{d}{dx}(uv) = u\,\frac{dv}{dx} + v\,\frac{du}{dx}.$$

(b) Quotient

$$\frac{d}{dx}\left(\frac{u}{v}\right) = \frac{v\,\dfrac{du}{dx} - u\,\dfrac{dv}{dx}}{v^2}.$$

(c) Composite Function (Function of a Function)

If $y = f(u)$ and $u = g(x)$ then $\dfrac{dy}{dx} = \left(\dfrac{dy}{du}\right)\left(\dfrac{du}{dx}\right).$

This is known as the chain rule.

(d) Standard Results

$y = f(x)$	$\dfrac{dy}{dx} = f'(x)$
x^n	nx^{n-1}
e^x	e^x
$a^x\,(a > 0)$	$a^x \ln a$
$\ln x$	$\dfrac{1}{x}$
$\log_a x$	$\dfrac{1}{x \ln a}$
$\sin x$	$\cos x$
$\cos x$	$-\sin x$
$\tan x$	$\sec^2 x$
$\operatorname{cosec} x$	$-\operatorname{cosec} x \cot x$
$\sec x$	$\sec x \tan x$
$\cot x$	$-\operatorname{cosec}^2 x$
$\sin^{-1} x$	$\dfrac{1}{\sqrt{(1 - x^2)}}$
$\cos^{-1} x$	$\dfrac{-1}{\sqrt{(1 - x^2)}}$
$\tan^{-1} x$	$\dfrac{1}{1 + x^2}$

14.2 Worked Examples

14.1 Define the derivative $f'(a)$ of a given function f at $x = a$.
Find from first principles:

(a) $f'(a)$ when $f(x) = \dfrac{3}{x}$ $\quad (x \neq 0)$;

(b) $h'(a)$ when $h(x) = x^3$;

(c) $g'(a)$ when $g(x) = 2x\,|x| + 1$.

• $f'(a) = \lim\limits_{b \to a} \left[\dfrac{f(a) - f(b)}{a - b}\right]$.

(a) $f(x) = \dfrac{3}{x}$; $f(a) - f(b) = \dfrac{3}{a} - \dfrac{3}{b} = \dfrac{3(b - a)}{ab}$.

$\dfrac{f(a) - f(b)}{a - b} = \dfrac{3(b - a)}{ab(a - b)} = \dfrac{-3}{ab}$,

$f'(a) = \lim\limits_{b \to a} \left(\dfrac{-3}{ab}\right) = \dfrac{-3}{a^2}$.

(b) $h(x) = x^3$; $h(a) - h(b) = a^3 - b^3 = (a - b)(a^2 + ab + b^2)$.

Therefore $\quad h'(a) = \lim\limits_{b \to a} \left[\dfrac{(a - b)(a^2 + ab + b^2)}{a - b}\right]$

$= \lim\limits_{b \to a} (a^2 + ab + b^2) = 3a^2$.

(c) $g(x) = 2x\,|x| + 1$.

    Consider in two parts: (i) $x \geqslant 0 \quad \Rightarrow \quad g(x) = 2x^2 + 1$;

                              (ii) $x < 0 \quad \Rightarrow \quad g(x) = -2x^2 + 1$.

    (i) $g(x) = 2x^2 + 1$.

$$g(a) - g(b) = (2a^2 + 1) - (2b^2 + 1) = 2(a - b)(a + b).$$

$$\frac{g(a) - g(b)}{a - b} = \frac{2(a - b)(a + b)}{a - b} = 2(a + b).$$

$$g'(a) = \lim_{b \to a}\;[2(a + b)] = 4a.$$

    (ii) $g(x) = -2x^2 + 1$.

$$g(a) - g(b) = (-2a^2 + 1) - (-2b^2 + 1) = -2(a - b)(a + b).$$

$$\frac{g(a) - g(b)}{a - b} = \frac{-2(a - b)(a + b)}{a - b} = -2(a + b).$$

$$g'(a) = \lim_{b \to a}\;[-2(a + b)] = -4a.$$

    So for $a \geqslant 0$, $g'(a) = 4a$, for $a < 0$, $g'(a) = -4a$.

    Combining these for all a, $g'(a) = 4\,|a|$.

14.2 By considering the limit of the gradient of the chord PQ of a parabola $y^2 = 4ax$, find the gradient of the tangent at P in terms of the parameter at P.

- The parametric equation of a parabola is $x = at^2$, $y = 2at$.
  At P, $t = p$; at Q, $t = q$.

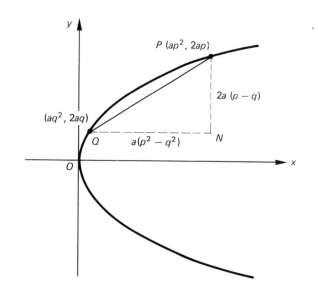

Gradient of chord PQ is: $\dfrac{2ap - 2aq}{ap^2 - aq^2}$

$$= \frac{2a(p - q)}{a(p - q)(p + q)} = \frac{2}{p + q}.$$

Gradient of the tangent at P is: $\displaystyle\lim_{q \to p} \left(\frac{2}{p + q} \right) = \frac{2}{2p} = \frac{1}{p}$.

14.3 Differentiate with respect to x:

(a) $x^3 \ln (3x)$;   (b) $\sin^4 3x$.

- (a) A product uv with $u = x^3$, so $\dfrac{du}{dx} = 3x^2$;

  $v = \ln (3x)$ (a composite function), so $\dfrac{dv}{dx} = \dfrac{1}{3x} \dfrac{d}{dx} (3x) = \dfrac{3}{3x} = \dfrac{1}{x}$.

  Thus
  $$\frac{d}{dx} [x^3 \ln (3x)] = (x^3)\left(\frac{1}{x}\right) + [\ln (3x)] (3x^2) = x^2 [1 + 3 \ln (3x)].$$

  (b) $\sin^4 3x$ is a composite function.
  Let $y = \sin^4 3x$ and $3x = u$.

  Then $y = \sin^4 u$, so $\dfrac{dy}{du} = (4 \sin^3 u) (\cos u)$.

  From $u = 3x$, $\dfrac{du}{dx} = 3$, hence $\dfrac{dy}{dx} = \dfrac{dy}{du} \dfrac{du}{dx} = (12 \sin^3 3x) (\cos 3x)$.

14.4 Differentiate

(a) $\ln \left(\dfrac{1 + x}{1 - x}\right)^{1/2}$,   (b) $\ln \tan^2 (4x + \pi)$.

- **With logarithmic functions it is sometimes easier to use the theory of logarithms to simplify the logarithm before differentiation.**

  (a) $\ln \left(\dfrac{1 + x}{1 - x}\right)^{1/2} = \tfrac{1}{2} \ln \left(\dfrac{1 + x}{1 - x}\right) = \tfrac{1}{2} [\ln (1 + x) - \ln (1 - x)]$.

  $$\frac{d}{dx} \{\tfrac{1}{2} [\ln (1 + x) - \ln (1 - x)]\} = \tfrac{1}{2} \left(\frac{1}{1 + x} - \frac{-1}{1 - x}\right)$$

  $$= \tfrac{1}{2} \left[\frac{(1 - x) + (1 + x)}{(1 + x) (1 - x)}\right] = \frac{1}{1 - x^2}.$$

  (b) $\ln [\tan^2 (4x + \pi)] = 2 \ln [\tan (4x + \pi)]$

  $$\frac{d}{dx} [2 \ln \tan (4x + \pi)] = 2 \frac{d}{dx} [\ln \tan (4x + \pi)].$$

  $y = \tan (4x + \pi)$ is a composite function.

  Let $u = 4x + \pi$, then $\dfrac{du}{dx} = 4$, $y = \tan u$, $\dfrac{dy}{du} = \sec^2 u$

  $$\frac{dy}{dx} = (\sec^2 u) (4) = 4 \sec^2 (4x + \pi).$$

  $$\frac{d}{dx} \{\ln [\tan^2 (4x + \pi)]\} = \frac{2 [4 \sec^2 (4x + \pi)]}{\tan (4x + \pi)} = \frac{8 \sec^2 (4x + \pi)}{\tan (4x + \pi)}$$

  $$= \frac{8}{\sin (4x + \pi) \cos (4x + \pi)}$$

  $$= \frac{16}{\sin (8x + 2\pi)}$$

  $$= \frac{16}{\sin 8x}.$$

14.5 Given that $y = \cos^{-1} x$, prove that $\dfrac{dy}{dx} = \dfrac{-1}{\sqrt{(1 - x^2)}}$.

● **With any inverse trigonometric function, change the original equation to $x = f(y)$ and find $\dfrac{dx}{dy}$. By the chain rule $\dfrac{dy}{dx} = 1 \div \dfrac{dx}{dy}$.**

$y = \cos^{-1} x \;\Rightarrow\; x = \cos y$, $\dfrac{dx}{dy} = -\sin y$, so $\dfrac{dy}{dx} = -\dfrac{1}{\sin y}$.

But $\sin y = \sqrt{(1 - \cos^2 y)} = \sqrt{(1 - x^2)}$.

Hence $\dfrac{dy}{dx} = \dfrac{-1}{\sqrt{(1 - x^2)}}$.

14.6 Differentiate with respect to x:

(a) $(2x - 3)^3 (x + 4)^{3/2}$;　(b) $\dfrac{(3x^2 + 5)^4}{(2x - 3)^3}$.

● (a) This is a product $y = uv$, where $u = (2x - 3)^3$, $v = (x + 4)^{3/2}$.
Using the chain rule,

$$\frac{du}{dx} = 6(2x - 3)^2 \qquad \text{and} \qquad \frac{dv}{dx} = \tfrac{3}{2}(x + 4)^{1/2}.$$

By the product rule,

$$\frac{d}{dx}\,[(2x - 3)^3 (x + 4)^{3/2}]$$

$$= (2x - 3)^3\,(\tfrac{3}{2})\,(x + 4)^{1/2} + 6(2x - 3)^2 (x + 4)^{3/2}.$$

Take out the common factor $\tfrac{3}{2}(2x - 3)^2 (x + 4)^{1/2}$:

$$\frac{dy}{dx} = \tfrac{3}{2}(2x - 3)^2 (x + 4)^{1/2}\,[(2x - 3) + 4(x + 4)]$$

$$= \tfrac{3}{2}(2x - 3)^2 (x + 4)^{1/2}\,(2x - 3 + 4x + 16)$$

$$= \tfrac{3}{2}(2x - 3)^2 (x + 4)^{1/2}\,(6x + 13).$$

(b) This is a quotient $y = \dfrac{u}{v}$ where

$$u = (3x^2 + 5)^4, \; v = (2x - 3)^3.$$

$$\frac{du}{dx} = 4(3x^2 + 5)^3 \frac{d}{dx}(3x^2 + 5) = 4(3x^2 + 5)^3 (6x) = 24x(3x^2 + 5)^3,$$

$$\frac{dv}{dx} = 3(2x - 3)^2 \frac{d}{dx}(2x - 3) = 3(2x - 3)^2 (2) = 6(2x - 3)^2.$$

Using the quotient rule,

$$\frac{dy}{dx} = \frac{v\,\dfrac{du}{dx} - u\,\dfrac{dv}{dx}}{v^2},$$

$$\frac{dy}{dx} = \frac{(2x - 3)^3\, 24x(3x^2 + 5)^3 - (3x^2 + 5)^4\, 6(2x - 3)^2}{(2x - 3)^6}.$$

155

The common factor in the numerator is $6 (2x - 3)^2 (3x^2 + 5)^3$;

$$\frac{dy}{dx} = \frac{6 (2x - 3)^2 (3x^2 + 5)^3 [4x (2x - 3) - (3x^2 + 5)]}{(2x - 3)^6}$$

$$= \frac{6 (3x^2 + 5)^3 (5x^2 - 12x - 5)}{(2x - 3)^4} .$$

14.7

(a) Differentiate $2x - \tan x + \frac{1}{3} \tan^3 x$, expressing your answer in terms of $\tan x$.

(b) Given that $y = ae^{-mx} \cos px$, prove that

$$\frac{d^2 y}{dx^2} + 2m \frac{dy}{dx} + (m^2 + p^2)y = 0.$$

(c) Given that $y = \ln (1 + 2x) - 2x + 2x^2$, show that

$$\frac{dy}{dx} \geqslant 0 \qquad \text{for all } x > -\tfrac{1}{2} .$$

• (a) Let $y = 2x - \tan x + \frac{1}{3} \tan^3 x$.

$$\frac{dy}{dx} = 2 - \sec^2 x + \tfrac{3}{3} \tan^2 x \sec^2 x$$

$$= 2 - 1 - \tan^2 x + \tan^2 x (1 + \tan^2 x)$$

$$= 1 + \tan^4 x.$$

(b) $y = ae^{-mx} \cos px$.

Method 1:

$$\frac{dy}{dx} = a (-me^{-mx} \cos px - pe^{-mx} \sin px)$$

$$= -my - ape^{-mx} \sin px. \qquad (1)$$

Differentiating again:

$$\frac{d^2 y}{dx^2} = -m \frac{dy}{dx} - ap (-me^{-mx} \sin px + pe^{-mx} \cos px). \qquad (2)$$

But, from (1), $ap\, e^{-mx} \sin px = -my - \dfrac{dy}{dx} .$

So equation 2 may be written:

$$\frac{d^2 y}{dx^2} = -m \frac{dy}{dx} - m^2 y - m \frac{dy}{dx} - p^2 y;$$

$$\frac{d^2 y}{dx^2} + 2m \frac{dy}{dx} + (m^2 + p^2)y = 0.$$

Method 2:
Multiply by e^{mx} to give $e^{mx} y = a \cos px.$
Differentiating with respect to x, once:

$$e^{mx} \frac{dy}{dx} + me^{mx} y = -ap \sin px,$$

and again:

$$e^{mx}\frac{d^2 y}{dx^2} + me^{mx}\frac{dy}{dx} + me^{mx}\frac{dy}{dx} + m^2 e^{mx}y = -ap^2\cos px.$$

Dividing by e^{mx} and writing $-ap^2 e^{-mx}\cos x = -p^2 y$ gives

$$\frac{d^2 y}{dx^2} + 2m\frac{dy}{dx} + (m^2 + p^2)y = 0.$$

(c) $y = \ln(1 + 2x) - 2x + 2x^2$.

$$\frac{dy}{dx} = \frac{2}{1 + 2x} - 2 + 4x$$
$$= \frac{8x^2}{1 + 2x}.$$

For all $x > -\frac{1}{2}$, $\dfrac{x^2}{1 + 2x} \geqslant 0$, hence $\dfrac{dy}{dx} \geqslant 0$.

14.8 Show, by considering $f'(x)$, that the function

$$f(x) = x^3 - 6x^2 + 13x - 3$$

is strictly increasing.
Find the gradient of the inverse function $f^{-1}(x)$ when $x = 5$.

- $f(x) = x^3 - 6x^2 + 13x - 3$.

$$f'(x) = 3x^2 - 12x + 13$$
$$= 3(x^2 - 4x + \tfrac{13}{3})$$
$$= 3[(x - 2)^2 + \tfrac{1}{3}].$$

The minimum value of $f'(x)$ is 1 when $x = 2$.
Therefore the gradient of the curve $y = f(x)$ is always positive, and the function is strictly increasing.

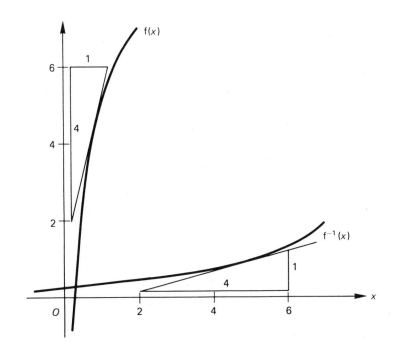

Since $f^{-1}(x)$ is difficult to compute, use the property of a function, namely that the graph of the inverse function $f^{-1}(x)$ is the result of reflecting the graph of $f(x)$ in the line $y = x$.

The gradient of the inverse function at $x = 5$ is

$$1/(\text{gradient of } f(x) \text{ when } f(x) = 5).$$

When $f(x) = 5$ then $\quad x^3 - 6x^2 + 13x - 3 = 5.$
$\qquad\qquad\Rightarrow\quad x^3 - 6x^2 + 13x - 8 = 0.$

By the factor theorem, one root is $x = 1$.
Since $f(x)$ is strictly increasing this is the only (real) root.
$\quad$ Thus $f(1) = 5$, $f'(1) = 4$ and the gradient of $f^{-1}(x)$ when $x = 5$ is $\frac{1}{4}$.

14.9 Given that the function f defined by

$$f(x) = 3 + x - x^2 \qquad \text{if } x \in (-\infty, 1)$$
$$\quad\;\; = x^2 + ax + b \qquad \text{if } x \in [1, \infty).$$

is continuous and has a continuous derivative for all values of x, show that $a = -3$ and $b = 5$.
$\quad$ Find the stationary points of f and sketch the graphs of f and its derivative f' in one diagram.

• For $f(x) = 3 + x - x^2$, $\quad x \in (-\infty, 1)$,

$$f(x) \to 3 + 1 - 1 \qquad \text{as } x \to 1.$$

$\quad$ For $f(x) = x^2 + ax + b$, $\quad x \in [1, \infty)$

$$f(x) = 1 + a + b \qquad \text{when } x = 1.$$

$\quad$ For a continuous function $3 = 1 + a + b$,
$$\Rightarrow\quad a + b = 2. \qquad\qquad (1)$$
$\quad$ For $f'(x) = 1 - 2x$, $\quad x \in (-\infty, 1)$,
$$f'(x) \to 1 - 2 \text{ as } x \to 1.$$

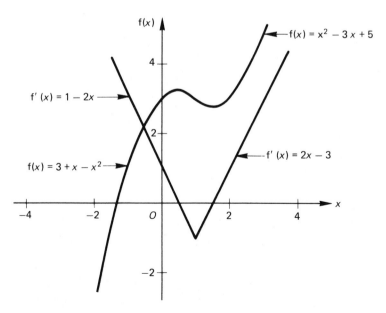

For $f'(x) = 2x + a$, $x \in [1, \infty)$,
$f'(x) = 2 + a$ when $x = 1$.
For a continuous derivative, $-1 = 2 + a$, so $a = -3$.
Substituting into equation 1 gives $b = 5$.
For $f(x) = 3 + x - x^2$, $x \in (-\infty, 1)$, $f'(x) = 1 - 2x$ and
$f''(x) = -2$.
$f'(x) = 0$ when $x = \frac{1}{2}$ and $f''(\frac{1}{2})$ is negative. Hence
$f(x)$ has a maximum point at $x = \frac{1}{2}$, $f(\frac{1}{2}) = (\frac{13}{4})$.
For $f(x) = x^2 - 3x + 5$, $x \in [1, \infty)$, $f'(x) = 2x - 3$ and
$f''(x) = 2$.
$f'(x) = 0$ when $x = \frac{3}{2}$ and $f''(\frac{3}{2})$ is positive. Hence
$f(x)$ has a minimum point at $x = \frac{3}{2}$, $f(\frac{3}{2}) = \frac{11}{4}$.

14.3 Exercises

14.1 Given that $f(x) = \dfrac{2}{x^2}$, $f'(x)$ equals

A, $\dfrac{-2}{x}$; B, $\dfrac{6}{x^3}$; C, $-\dfrac{1}{x}$; D, $-\dfrac{4}{x^3}$; E, none of these.

14.2 Given that $f(x) = (2x + 1)^3$, $f'(2)$ equals
   A, 50; B, 75; C, 625/8; D, 150; E, none of these.

14.3 Differentiate with respect to x:
(a) $\log_e [\tan^3 (4x + 5)]$ $(0 < 4x + 5 < \pi/2)$;
(b) $\dfrac{x^3}{(2x + 1)}$;
(c) $e^{3x} \sin 2x$.

14.4
(a) Differentiate $(x - 2)^3 e^{-5x}$ with respect to x.

(b) If $y = \cos 2x + 2 \sin x$, where $0 \leqslant x \leqslant \pi$, find the values of x for which $\dfrac{dy}{dx} = 0$.

Sketch the graph of y against x for $0 \leqslant x \leqslant \pi$.

14.5 If $y = \left(\dfrac{1 - 3x^2}{2 + 5x^2} \right)^{1/2}$, find the values of A and B which satisfy

$\dfrac{dy}{dx} = y \left(\dfrac{Ax}{1 - 3x^2} + \dfrac{Bx}{2 + 5x^2} \right)$.

14.6
(a) Evaluate $\lim\limits_{x \to 0} \left(\dfrac{\cos 4x - \cos 2x}{\cos 3x - \cos x} \right)$.

(b) Assuming that $\lim\limits_{x \to 0} \dfrac{\sin x}{x} = 1$, evaluate the limits:

   (i) $\lim\limits_{x \to 0} \left(\dfrac{1 - \cos 2x}{x^2} \right)$,   (ii) $\lim\limits_{x \to \pi/2} \left(\dfrac{2x - \pi}{\cos x} \right)$.

14.7 The function f defined by

$$f(x) = ax^2 + 8x + 2 \qquad \text{if } x \in (-\infty, 1)$$
$$= b(2x - 3)^2 \qquad \text{if } x \in [1, \infty)$$

is continuous and has a continuous derivative for all values of x. Find the values of the constants a and b. Find the stationary points and sketch the graphs of f and f′ in the neighbourhood of $x = 1$.

14.8 Find the gradient of the chord PQ of the curve $x = at^3$, $y = at^2$, where P and Q have parameters p and q respectively. Use the gradient of the chord to find the gradient of the tangent at Q.

14.4 Outline Solutions to Exercises

14.1 $f(x) = \dfrac{2}{x^2} = 2x^{-2}$

$\Rightarrow \quad f'(x) = -4x^{-3} = \dfrac{-4}{x^3}.$ Answer D

14.2 $f(x) = (2x + 1)^3$, $f'(x) = 3(2)(2x + 1)^2$,
$\Rightarrow \quad f'(2) = 6(4 + 1)^2 = 150.$ Answer D

14.3
(a) $\log_e [\tan^3 (4x + 5)] = 3 \log_e [\tan (4x + 5)].$

$$\frac{dy}{dx} = \frac{[3 \sec^2 (4x + 5)] (4)}{\tan (4x + 5)} = \frac{12}{\sin (4x + 5) \cos (4x + 5)}$$

$$= \frac{24}{\sin (8x + 10)}.$$

(b) Quotient rule for $y = \dfrac{x^3}{(2x + 1)}$:

$$\frac{dy}{dx} = \frac{(2x + 1)(3x^2) - (x^3)(2)}{(2x + 1)^2} = \frac{x^2 (4x + 3)}{(2x + 1)^2}.$$

(c) Product rule for $y = e^{3x} \sin 2x$:

$$\frac{dy}{dx} = (e^{3x})(2 \cos 2x) + (3e^{3x})(\sin 2x)$$

$$= e^{3x} (2 \cos 2x + 3 \sin 2x).$$

14.4
(a) Product rule for $y = (x - 2)^3 e^{-5x}$:

$$\frac{dy}{dx} = (x - 2)^3 (-5e^{-5x}) + (e^{-5x}) [3(x - 2)^2]$$

$$= e^{-5x} (x - 2)^2 (13 - 5x).$$

(b) If $y = \cos 2x + 2 \sin x$, then $\dfrac{dy}{dx} = -2 \sin 2x + 2 \cos x$;

$\dfrac{dy}{dx} = 0$ when $-4 \sin x \cos x + 2 \cos x = 0.$

$$\cos x = 0 \quad \text{or} \quad \sin x = 0.5 \quad \Rightarrow \quad x = \frac{\pi}{6}, \frac{\pi}{2} \text{ or } \frac{5\pi}{6}.$$

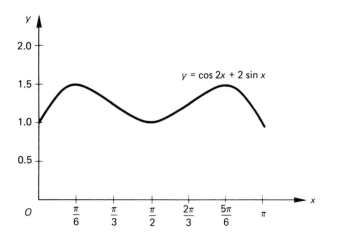

14.5 Take logs to base e before differentiating:

$$\ln y = \ln \sqrt{\left(\frac{1 - 3x^2}{2 + 5x^2}\right)}$$

$$\Rightarrow \quad \ln y = \tfrac{1}{2} \ln (1 - 3x^2) - \tfrac{1}{2} \ln (2 + 5x^2).$$

$$\frac{1}{y}\frac{dy}{dx} = \frac{-6x}{2(1 - 3x^2)} - \frac{10x}{2(2 + 5x^2)}.$$

$$\frac{dy}{dx} = y\left(\frac{-3x}{(1 - 3x^2)} - \frac{5x}{(2 + 5x^2)}\right).$$

Hence $A = -3, B = -5$.

14.6

(a)
$$\frac{\cos 4x - \cos 2x}{\cos 3x - \cos x} = \frac{-2 \sin 3x \sin x}{-2 \sin 2x \sin x} = \frac{\sin 3x}{\sin 2x}.$$

Since $\lim_{x \to 0} \dfrac{\sin ax}{x} = a \quad \Rightarrow \quad \lim_{x \to 0}\left(\dfrac{\sin 3x}{\sin 2x}\right) = \lim_{x \to 0}\left[\left(\dfrac{\sin 3x}{x}\right)\left(\dfrac{x}{\sin 2x}\right)\right] = \dfrac{3}{2}.$

(b) (i)
$$\frac{1 - \cos 2x}{x^2} = \frac{2 \sin^2 x}{x^2} = 2\left(\frac{\sin x}{x}\right)^2,$$

$$\lim_{x \to 0} 2\left(\frac{\sin x}{x}\right)^2 = 2.$$

(ii) Let $y = \dfrac{\pi}{2} - x$, then $\cos x = \sin y$ and $2x - \pi = -2y$

$$\Rightarrow \quad \lim_{x \to \pi/2}\left(\frac{2x - \pi}{\cos x}\right) = \lim_{y \to 0}\frac{(-2y)}{\sin y}$$

$$= \lim_{y \to 0} -2\left(\frac{y}{\sin y}\right)$$

$$= -2.$$

14.7 $f(x) \begin{cases} = ax^2 + 8x + 2 \\ = b(2x-3)^2 \end{cases}$ · $f'(x) \begin{cases} = 2ax + 8 & x \in (-\infty, 1) \\ = 4b(2x-3) & x \in [1, \infty). \end{cases}$

f and f′ must be continuous at $x = 1$,

⇒ $a + 10 = b$ (1)   and $2a + 8 = -4b$ (2).

From equations 1 and 2, $a = -8$ and $b = 2$.

When $f'(x) = 2ax + 8 = -16x + 8$, stationary point is at $x = \frac{1}{2}$, $f(x) = 4$.

When $f'(x) = 4b(2x - 3) = 8(2x - 3)$, stationary point is at $x = \frac{3}{2}$, $f(x) = 0$.

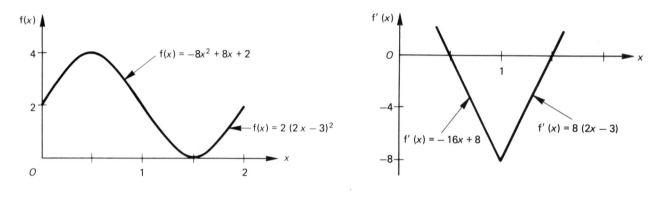

14.8 $P(ap^3, ap^2)$ and $Q(aq^3, aq^2)$.

Gradient $PQ = \dfrac{a(p^2 - q^2)}{a(p^3 - q^3)} = \dfrac{p + q}{p^2 + pq + q^2}$.

Gradient of tangent at Q is $\lim\limits_{p \to q} \dfrac{p + q}{p^2 + pq + q^2} = \dfrac{2}{3q}$.

15 Differentiation—2

Differentiation of simple functions defined implicitly or parametrically.
Application of differentiation to gradients, tangents and normals, maxima and minima (and points of inflexion), curve sketching, connected rates of change, small increments and approximations.

15.1 Fact Sheet

(a) Implicit Functions

$$\frac{d}{dx}(f(y)) = (f'(y))\left(\frac{dy}{dx}\right) \text{ where } f'(y) = \frac{d}{dy}(f(y)).$$

(b) Parametric Functions

If $x = x(t)$ and $y = y(t)$ then $\dfrac{dy}{dx} = \dfrac{\frac{dy}{dt}}{\frac{dx}{dt}} = \dfrac{\dot{y}}{\dot{x}}$

where $\dot{x} = \dfrac{dx}{dt}$ and $\dot{y} = \dfrac{dy}{dt}$.

(c) Tangents and Normals at (x_0, y_0)

If $\dfrac{dy}{dx} = m$ when evaluated at (x_0, y_0) then:

(i) the tangent at (x_0, y_0) may be written as
$$y - y_0 = m(x - x_0),$$

(ii) the normal at (x_0, y_0) may be written as
$$y - y_0 = \frac{-1}{m}(x - x_0).$$

(d) Maxima and Minima

Stationary points occur when $\dfrac{dy}{dx} = 0$.

At such points, if

(i) $\dfrac{d^2y}{dx^2} < 0$ the point is a maximum,

(ii) $\dfrac{d^2y}{dx^2} > 0$ the point is a minimum,

(iii) $\dfrac{d^2y}{dx^2} = 0$ the point can be maximum, minimum or a point of inflexion.

(e) Points of Inflexion

If $\dfrac{d^2y}{dx^2} = 0$ at $x = x_0$, and there is a change of sign of $\dfrac{d^2y}{dx^2}$ as x passes through x_0,

then it is a point of inflexion. This is independent of the value of $\dfrac{dy}{dx}$ at $x = x_0$.

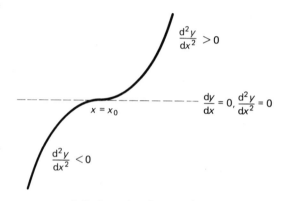

Inflexion and stationary point

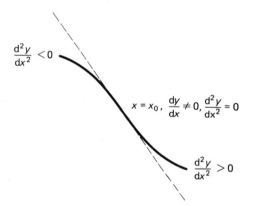

Inflexion but not a stationary point

(f) Rates of Change

$f'(x)$ or $\dfrac{dy}{dx}$ represents the rate of increase of a function $f(x)$ or y with respect to x.

The rate of decrease of $f(x)$ or y is $- f'(x)$ or $- \dfrac{dy}{dx}$.

(g) Connected Rates of Change

If $y = y(x)$ and the rate of change of x with respect to t is $\dfrac{dx}{dt}$ then the rate of change of y with respect to t is $\dfrac{dy}{dt}$ given by $\dfrac{dy}{dt} = \left(\dfrac{dy}{dx}\right)\left(\dfrac{dx}{dt}\right)$.

(h) Small Increments

If $y = f(x)$ then $\dfrac{\delta y}{\delta x} \approx f'(x) \quad \Rightarrow \quad \delta y \approx f'(x)\,\delta x$.

(i) Velocity and Acceleration

If $x = x(t)$ then velocity $v = \dfrac{dx}{dt} = \dot{x}$,

acceleration $a = \dfrac{dv}{dt} = \dfrac{d^2 x}{dt^2} = \ddot{x} = v\,\dfrac{dv}{dx} = \dfrac{1}{2}\dfrac{d}{dx}(v^2)$.

15.2 Worked Examples

15.1 Show that the equations of the tangent and normal to the curve $y^2 = 4ax$ at the point $P\,(at^2,\ 2at)$ are $ty = x + at^2$ and $y + tx = 2at + at^3$.

- At the point P, $x = at^2$, $y = 2at$.

  Differentiating, $\dfrac{dx}{dt} = 2at$, $\dfrac{dy}{dt} = 2a$.

  But $\quad \dfrac{dy}{dx} = \left(\dfrac{dy}{dt}\right)\left(\dfrac{dt}{dx}\right) = \left(\dfrac{dy}{dt}\right) \div \left(\dfrac{dx}{dt}\right) \quad \Rightarrow \quad \dfrac{dy}{dx} = \dfrac{2a}{2at} = \dfrac{1}{t}$.

  Therefore the gradient of the tangent is $\dfrac{1}{t}$ and the gradient of the normal is $-t$.

  Equation of the tangent at $(at^2,\ 2at)$ is

  $$y - 2at = \frac{x - at^2}{t} \quad \Rightarrow \quad ty = x + at^2.$$

  Equation of the normal at $(at^2,\ 2at)$ is

  $$y - 2at = -t(x - at^2) \quad \Rightarrow \quad y + tx = 2at + at^3.$$

15.2 The radius of a sphere is decreasing at 3 cm s^{-1}. Obtain the rate of decrease of the surface area of the sphere when the radius is 18 cm. Leave your answer in terms of π.

- Given that the rate of **decrease** of the radius is 3 cm s^{-1} then $\dfrac{dr}{dt} = -3$.

Surface area of a sphere $A = 4\pi r^2$, $\dfrac{dA}{dr} = 8\pi r$.

The rate of **increase** of the surface area is $\dfrac{dA}{dt}$.

$$\frac{dA}{dt} = \left(\frac{dA}{dr}\right)\left(\frac{dr}{dt}\right) = (8\pi r)\,(-3) = -24\pi r.$$

Therefore the rate of **decrease** of the surface area is $24\pi r$.
When $r = 18$, the rate of decrease of surface area is 432π cm^2 s^{-1}.

15.3

(a) Given that $y = \dfrac{\cos x + \sin x}{\cos x - \sin x}$, show that $\dfrac{d^2 y}{dx^2} = 2y\,\dfrac{dy}{dx}$

(b) Find $\dfrac{dy}{dx}$ when (i) $y = \ln (1 + x^3)^2$,

(ii) $x = \sin^3 \theta$, $y = \cos 2\theta$.

● (a) $y = \dfrac{\cos x + \sin x}{\cos x - \sin x}$.

Use the quotient rule.

$$\frac{dy}{dx} = \frac{(\cos x - \sin x)(-\sin x + \cos x) - (\cos x + \sin x)(-\sin x - \cos x)}{(\cos x - \sin x)^2}$$

$$= \frac{(\cos x - \sin x)^2 + (\cos x + \sin x)^2}{(\cos x - \sin x)^2} = 1 + y^2.$$

Differentiate implicitly:

Remember: $\dfrac{d(y^2)}{dx} = (2y)\left(\dfrac{dy}{dx}\right)$.

$$\frac{d^2 y}{dx^2} = \frac{d}{dx}(1) + \frac{d}{dx}(y^2).$$

Hence $\dfrac{d^2 y}{dx^2} = (2y)\left(\dfrac{dy}{dx}\right)$

(b) (i) $y = \ln (1 + x^3)^2 = 2\ln(1 + x^3)$.
   Use the chain rule:
$$\frac{dy}{dx} = (2)\left(\frac{1}{1 + x^3}\right)(3x^2) = \frac{6x^2}{1 + x^3}.$$

(ii) Parametric differentiation:
$$\frac{dy}{dx} = \left(\frac{dy}{d\theta}\right)\left(\frac{d\theta}{dx}\right).$$

$x = \sin^3 \theta$, $\dfrac{dx}{d\theta} = 3\sin^2 \theta \cos \theta$; $y = \cos 2\theta$, $\dfrac{dy}{d\theta} = -2\sin 2\theta$:

$$\Rightarrow \frac{dy}{dx} = \frac{-2\sin 2\theta}{3\sin^2 \theta \cos \theta} = \frac{-4\sin \theta \cos \theta}{3\sin^2 \theta \cos \theta} = \frac{-4\,\mathrm{cosec}\,\theta}{3}.$$

15.4 A body starts from O and moves in a straight line. Its distance from O after t seconds is s metres, where $s = 4te^{-t^2/3}$. Find the velocity of the body and determine when it first comes instantaneously to rest. Find the acceleration of the body at this time. Show that it subsequently moves towards O without ever reaching it. Give numerical answers to three significant figures.

- Given that $s = 4te^{-t^2/3}$ (1)

  velocity of the body $v = \dfrac{ds}{dt} = 4e^{-t^2/3} + 4t\left(-\tfrac{2}{3}te^{-t^2/3}\right)$

  $$= \tfrac{4}{3}e^{-t^2/3}(3 - 2t^2). \qquad (2)$$

  When $v = 0$, $e^{-t^2/3} = 0$ or $3 - 2t^2 = 0$.

  $e^{-t^2/3} \neq 0$ for finite t, therefore $2t^2 = 3$, $t = \sqrt{\tfrac{3}{2}} = 1.22$

  (taking the positive value only for time).

  Acceleration of the body $a = \dfrac{dv}{dt}$.

  $$\frac{dv}{dt} = \tfrac{4}{3}e^{-t^2/3}(-4t) + (3 - 2t^2)\tfrac{4}{3}\left(-\tfrac{2}{3}te^{-t^2/3}\right)$$

  $$= \tfrac{4}{3}e^{-t^2/3}\left(-4t - 2t + \tfrac{4}{3}t^3\right) = \tfrac{8}{9}te^{-t^2/3}(2t^2 - 9).$$

  When $t = \sqrt{\tfrac{3}{2}}$, $a = \tfrac{8}{9}\sqrt{\tfrac{3}{2}}e^{-1/2}(3 - 9) = -\tfrac{16}{3}\sqrt{\tfrac{3}{2}}e^{-1/2} = -3.96$.

  Therefore the body is instantaneously at rest after 1.22 seconds. The acceleration at this time is -3.96 m s^{-2}.

  The velocity now becomes negative and the particle moves towards O, and continues in that direction since v remains negative (from equation 2). But (from equation 1) s is always positive.

  Therefore after $t = 1.22$ the body moves towards O but never reaches it.

15.5 Find the set of real values of x which satisfy $(x^2 - 4)(x - 2) > 0$.

Find the coordinates of the maximum point, the minimum point and the point of inflexion of the curve $y = x^3 - 2x^2 - 4x + 8$.

Sketch this curve showing its intersections with the axes. Find also the set of real values of x which satisfy $(x^2 - 4)(x + 2) \leq 0$.   (L)

- Let $y = (x^2 - 4)(x - 2) = (x + 2)(x - 2)^2$,

  $(x - 2)^2 > 0$ for $x \neq 2$, $(x + 2) > 0$ for $x > -2$.

  Hence $(x^2 - 4)(x - 2) > 0$ for $x > -2$, $x \neq 2$.

  $y = x^3 - 2x^2 - 4x + 8 = (x^2 - 4)(x - 2) = (x + 2)(x - 2)^2$,

  $\dfrac{dy}{dx} = 3x^2 - 4x - 4 = (3x + 2)(x - 2)$,

  $\dfrac{dy}{dx} = 0$ when $x = 2$ or $-\tfrac{2}{3}$.

  $\dfrac{d^2y}{dx^2} = 6x - 4 \quad \Rightarrow \quad \dfrac{d^2y}{dx^2} = 0$ when $x = \tfrac{2}{3}$.

  When $x = 2$, $\dfrac{d^2y}{dx^2} = 8$ (positive) $\quad \Rightarrow \quad$ there is a minimum point at $(2, 0)$.

  When $x = -\tfrac{2}{3}$, $\dfrac{d^2y}{dx^2} = -8$ (negative) $\quad \Rightarrow \quad$ there is a maximum point at $(-\tfrac{2}{3}, \tfrac{256}{27})$.

  When $x = \tfrac{2}{3}$, $\dfrac{dy}{dx} \neq 0$, $\dfrac{d^2y}{dx^2} = 0$, $\quad \Rightarrow \quad$ inflexion point at $(\tfrac{2}{3}, \tfrac{128}{27})$.

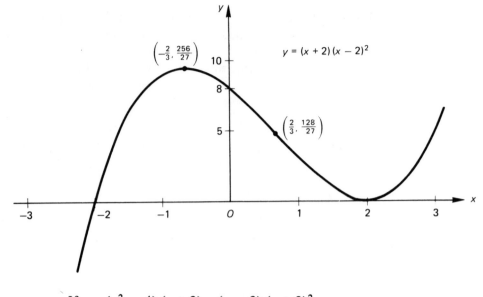

If $y = (x^2 - 4)(x + 2) = (x - 2)(x + 2)^2$,
$(x + 2)^2 \geqslant 0$ for all x, $(x - 2) \leqslant 0$ for $x \leqslant 2$,
$\Rightarrow \quad (x^2 - 4)(x + 2) \leqslant 0$ for $x \leqslant 2$.

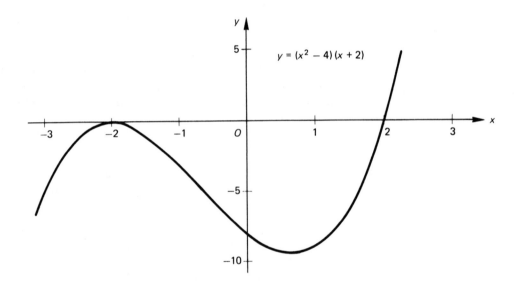

15.6 Find the coordinates of the maximum point T and the minimum point B of the curve $y = \dfrac{x^3}{3} - 2x^2 + 3x$.

Find also the point of inflexion I and show that T, I and B are collinear. Calculate to the nearest $0.1°$ the acute angle between TIB and the normal to the curve at I. (L)

- $y = \dfrac{x^3}{3} - 2x^2 + 3x.$

$\dfrac{dy}{dx} = x^2 - 4x + 3 \quad \Rightarrow \quad \dfrac{dy}{dx} = 0$ when $x = 3$ or 1.

$\dfrac{d^2y}{dx^2} = 2x - 4.$

When $x = 3$, $\dfrac{d^2 y}{dx^2}$ is positive, a minimum point.

When $x = 1$, $\dfrac{d^2 y}{dx^2}$ is negative, a maximum point.

When $x = 2$, $\dfrac{d^2 y}{dx^2} = 0$ and $\dfrac{dy}{dx} \neq 0$. Hence an inflexion point.

Thus T has coordinates $(1, \frac{4}{3})$, B $(3, 0)$ and I $(2, \frac{2}{3})$.
Gradient of TI is $(\frac{2}{3} - \frac{4}{3})/(2 - 1) = -\frac{2}{3}$.
Gradient of IB is $(0 - \frac{2}{3})/(3 - 2) = -\frac{2}{3}$.
Therefore T, I and B are collinear.

At I, $\dfrac{dy}{dx} = -1$, therefore gradient of normal $= 1$.

From the sketch: $\angle INB = 45°$.
Since the gradient of TIB is $-\frac{2}{3}$, $\angle IBN = \tan^{-1} \frac{2}{3} = 33.7°$.
Hence $\angle NIT = 78.7°$ (exterior angle of triangle INB).

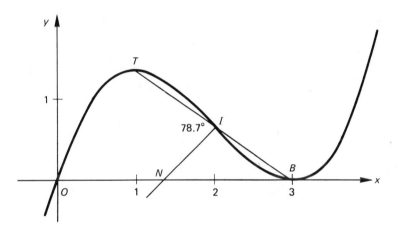

15.7 Given that $y = \ln (1 + \cos x)^2$, prove that $\dfrac{d^2 y}{dx^2} + 2e^{-y/2} = 0$.

What can be said about any stationary values of y?
Sketch the curve $y = \ln (1 + \cos x)^2$ for $-2\pi \leqslant x \leqslant 2\pi$.

● $y = \ln (1 + \cos x)^2 = 2 \ln (1 + \cos x)$.

Using the chain rule, $\dfrac{dy}{dx} = 2 \left(\dfrac{1}{1 + \cos x} \right) (-\sin x)$.

Using the quotient rule with $u = -2 \sin x$, $v = 1 + \cos x$,

$$\frac{du}{dx} = -2 \cos x, \quad \frac{dv}{dx} = -\sin x,$$

$$\frac{d^2 y}{dx^2} = \frac{(1 + \cos x)(-2 \cos x) + (2 \sin x)(-\sin x)}{(1 + \cos x)^2}$$

$$= \frac{-2 \cos x - 2}{(1 + \cos x)^2} = \frac{-2}{1 + \cos x}.$$

But $1 + \cos x = e^{y/2}$,

therefore $\dfrac{d^2 y}{dx^2} = -2e^{-y/2}$, $\Rightarrow$ $\dfrac{d^2 y}{dx^2} + 2e^{-y/2} = 0$.

Since $e^{-y/2}$ is positive for all finite values of y, $\dfrac{d^2 y}{dx^2} < 0$ for all finite

values of y. Hence any stationary values of y are maximum values.

Sketch:
(a) $\ln (1 + \cos x)^2$ is not defined when $\cos x = -1$
$\Rightarrow$ asymptotes are $x = \pm\pi$.
(b) $\dfrac{dy}{dx} = 0$ when $\sin x = 0$, but $1 + \cos x \neq 0$

$\Rightarrow \quad \dfrac{dy}{dx} = 0$ when $x = -2\pi, 0, 2\pi$.
(c) $\ln (1 + \cos x)^2$ is an even function.

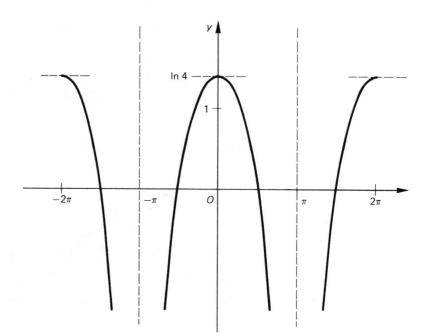

15.8 Find the stationary points of the function given by $f(x) = \dfrac{(x + 1)(x + 4)}{x}$

Sketch the graph of $f(x)$ and find the domain and range of $f(x)$.

- $f(x) = \dfrac{(x + 1)(x + 4)}{x} = \dfrac{x^2 + 5x + 4}{x}$.

By the quotient rule,

$$f'(x) = \frac{x(2x + 5) - (x^2 + 5x + 4)}{x^2}$$

$$= \frac{x^2 - 4}{x^2}.$$

$f'(x) = 0$ when $x^2 - 4 = 0 \Rightarrow x = \pm 2$.
Hence stationary points are at $(2, 9)$ and $(-2, 1)$.
Sketch:
(a) $f(x)$ is not defined when $x = 0 \Rightarrow$ asymptote is $x = 0$.
(b) $f(x) = 0$ when $x = -1$ or -4.
(c) $f(x) = \dfrac{x^2 + 5x + 4}{x} = x + 5 + \dfrac{4}{x}$.
When $|x|$ is large, $f(x) \to x + 5$ (asymptote).

170

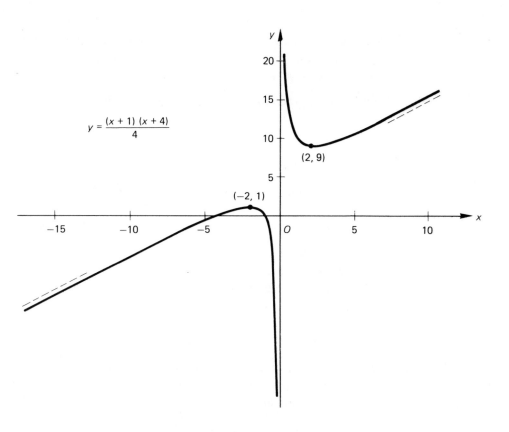

$$y = \frac{(x+1)(x+4)}{4}$$

$(2, 9)$

$(-2, 1)$

(d) When x is small, $f(x) \to \dfrac{4}{x}$.

The domain of $f(x)$ is $x \in \mathbb{R}$, $x \neq 0$.

The range of $f(x)$ is $f(x) \leqslant 1$ or $f(x) \geqslant 9$.

15.3 Exercises

15.1 Show that $e^{-x}\cos x$ has turning values when $x = (n + \frac{3}{4})\pi$ when n is an integer. Distinguish between maxima and minima.

Sketch the curve $y = e^{-x} \cos x$ for $0 \leqslant x \leqslant 2\pi$.

15.2

(a) A curve is given parametrically by $x = t^2 - 2t$, $y = t^2 + 2t$. Find $\dfrac{dy}{dx}$ and $\dfrac{d^2 y}{dx^2}$

in terms of t and hence find the coordinates of the stationary point on the curve, determining its nature.

Sketch the curve for $-2 \leqslant t \leqslant 2$.

(b) If $y = \sin^2 x \cos^3 x$, find the values of x lying between $-\pi/2$ and π for which $\dfrac{dy}{dx} = 0$.

Sketch the curve $y = \sin^2 x \cos^3 x$.

15.3 Given that $y = e^{-4x} \cos 3x$,

(a) express $\dfrac{d^2 y}{dx^2}$ in the form $Ae^{-4x} [\sin (3x + \alpha)]$, giving the values of A and $\tan \alpha$.

(b) Prove that $\dfrac{d^2 y}{dx^2} + 8 \dfrac{dy}{dx} + 25y = 0$.

15.4 A point moves along the x-axis so that, at time t, its displacement from the origin is given by $x = A \sin wt$, where A and w are constants. Prove that if, at time t, the velocity is v and the acceleration is a, then (a) $v^2 = w^2 (A^2 - x^2)$, (b) $a = -w^2 x$, and (c) the particle oscillates with period $\dfrac{2\pi}{w}$.

15.5

(a) (i) Find any stationary points and points of inflexion of $y = \dfrac{\ln x}{x}$.

    (ii) Sketch the graph of $y = \dfrac{\ln x}{x}$, for $x > 0$, indicating clearly any of the points found in (i).

(b) Find the gradient of the curve $x^3 + 3x^2 y + y^3 + 5 = 0$ at the point $(2, -1)$. Write down the equations of the tangent and normal at this point.

15.6 Find the stationary points of the function f where $f(x) = \dfrac{4x - 5}{(x - 1)(x + 1)}$ and determine the nature of each point.

Sketch the graph of f and give the equations of the asymptotes.
Give the domain and range of f.

15.7 A solid cylinder, of height h and base radius r, has a fixed volume V. Find the ratio $r : h$ if the surface area of the cylinder is a minimum.

15.8 The equation of a curve is given parametrically by $x = 2(\theta - \sin \theta)$, $y = 2(1 - \cos \theta)$. Show that $\dfrac{dy}{dx} = \cot \dfrac{\theta}{2}$.

At A, $\theta = \dfrac{\pi}{2}$, and at B, $\theta = \dfrac{3\pi}{2}$. Find the equations of the tangents at A and B.

15.4 Outline Solutions to Exercises

15.1 $y' = e^{-x}(-\cos x - \sin x)$; $y' = 0$ when $\cos x + \sin x = 0$
$\Rightarrow \quad \tan x = -1 \quad \Rightarrow \quad x = (n + \tfrac{3}{4})\pi$.
$y'' = e^{-x}(\cos x + \sin x + \sin x - \cos x) = 2e^{-x} \sin x$.
When n is even, $\sin(n + \tfrac{3}{4})\pi > 0 \quad \Rightarrow \quad y$ has a minimum value.
When n is odd, $\sin(n + \tfrac{3}{4})\pi < 0 \quad \Rightarrow \quad y$ has a maximum value.

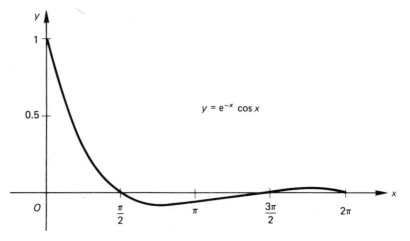

$y = e^{-x} \cos x$

15.2

(a) $\dfrac{dx}{dt} = 2t - 2,$ $\dfrac{dy}{dt} = 2t + 2,$ $\Rightarrow$ $\dfrac{dy}{dx} = \dfrac{t+1}{t-1}.$

$$\dfrac{d^2 y}{dx^2} = \dfrac{d}{dx}\left(\dfrac{dy}{dx}\right) = \left[\dfrac{d}{dt}\left(\dfrac{dy}{dx}\right)\right]\left(\dfrac{dt}{dx}\right) = \dfrac{(t-1)-(t+1)}{(t-1)^2} \cdot \dfrac{1}{2(t-1)}$$

$$= \dfrac{-1}{(t-1)^3}.$$

$\dfrac{dy}{dx} = 0$ when $t = -1$, and then $\dfrac{d^2 y}{dx^2} = \dfrac{1}{8}.$

Hence the point $(3, -1)$ is a minimum.

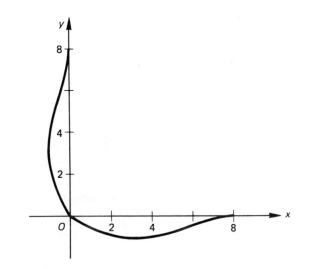

(b) $y' = (2 \sin x \cos x) \cos^3 x + \sin^2 x\, (-3 \cos^2 x \sin x)$
$\qquad = \cos^2 x \sin x\, (2 \cos^2 x - 3 \sin^2 x).$

$y' = 0$ when $\cos x = 0$ $\Rightarrow$ $x = \pm \dfrac{\pi}{2}$; or $\sin x = 0$ $\Rightarrow$ $x = 0$ or π;

$\qquad$ or $\tan x = \pm\sqrt{\tfrac{2}{3}}$ $\Rightarrow$ $x = -0.68, 0.68,$ or 2.46 radians.

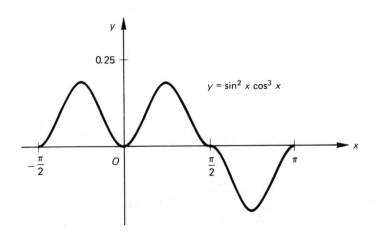

15.3

(a) $y' = e^{-4x} (-3 \sin 3x - 4 \cos 3x)$

    $y'' = e^{-4x} (24 \sin 3x + 7 \cos 3x)$

       $= 25e^{-4x} \sin (3x + \alpha)$ where $\tan \alpha = \frac{7}{24}$.

(b) Multiply by e^{4x}, $e^{4x} y = \cos 3x$.

    $e^{4x} y' + 4e^{4x} y = -3 \sin 3x,$

    $e^{4x} y'' + 8e^{4x} y' + 16e^{4x} y = -9 \cos 3x = -9e^{4x} y.$

    Divide by e^{4x}, $y'' + 8y' + 25y = 0$.

15.4 $v = \dot{x} = wA \cos wt,\quad a = \ddot{x} = -w^2 A \sin wt = -w^2 x,$
$\cos^2 wt = 1 - \sin^2 wt,\quad$ hence $v^2 = w^2 (A^2 - x^2)$.

Period T of $\sin wt$ is given by $wT = 2\pi \quad \Rightarrow \quad T = \dfrac{2\pi}{w}$.

15.5

(a) $y' = \left(-\dfrac{1}{x^2}\right) (\ln x) + \left(\dfrac{1}{x}\right)\left(\dfrac{1}{x}\right) = \dfrac{1}{x^2} (1 - \ln x)$.

    $y' = 0$ when $\ln x = 1 \quad \Rightarrow \quad x = e$.

    $y'' = \dfrac{-2}{x^3} (1 - \ln x) + \dfrac{1}{x^2}\left(\dfrac{-1}{x}\right) = \dfrac{-1}{x^3} (3 - 2 \ln x);$

    $y'' = 0$ when $\ln x = \frac{3}{2} \quad \Rightarrow \quad x = e^{3/2}$ (inflexion).

    When $x = e$, $y'' = \dfrac{-1}{e^3}$ (maximum),

  $\Rightarrow \quad$ maximum at (e, e^{-1}), inflexion at $\left(e^{3/2}, \dfrac{3e^{-3/2}}{2}\right)$.

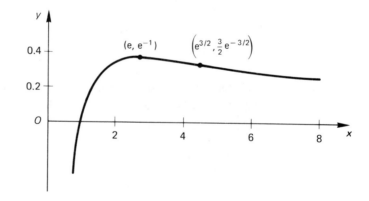

(b) $3x^2 + 3x^2 y' + 6xy + 3y^2 y' = 0$

  $\Rightarrow \quad y' = \dfrac{-(x^2 + 2xy)}{x^2 + y^2}$. At $(2, -1)$, $y' = 0$.

    Equation of tangent: $y = -1$; equation of normal: $x = 2$.

15.6 $f'(x) = \dfrac{(x^2 - 1)(4) - (4x - 5)(2x)}{(x - 1)^2 (x + 1)^2}$

       $= \dfrac{-2(2x - 1)(x - 2)}{(x - 1)^2 (x + 1)^2}$.

  $f'(x) = 0$ when $x = \frac{1}{2}$ or $x = 2$.

When $x < \frac{1}{2}$, f'(x) is negative, ($x \neq -1$);
when $\frac{1}{2} < x < 2$, f'(x) is positive, ($x \neq 1$);
when $x > 2$, f'(x) is negative.

  Hence at $x = \frac{1}{2}$, f has a minimum point ($\frac{1}{2}$, 4),

        at $x = 2$, f has a maximum point (2, 1).

  Sketch: (a) asymptotes at $x = \pm 1$;

              (b) f(x) = 0 when $x = \frac{5}{4}$;

              (c) as $|x| \to \infty$, f(x) $\to \dfrac{4}{x}$ $\Rightarrow$ $y = 0$ is an asymptote;

              (d) min ($\frac{1}{2}$, 4), max (2, 1).

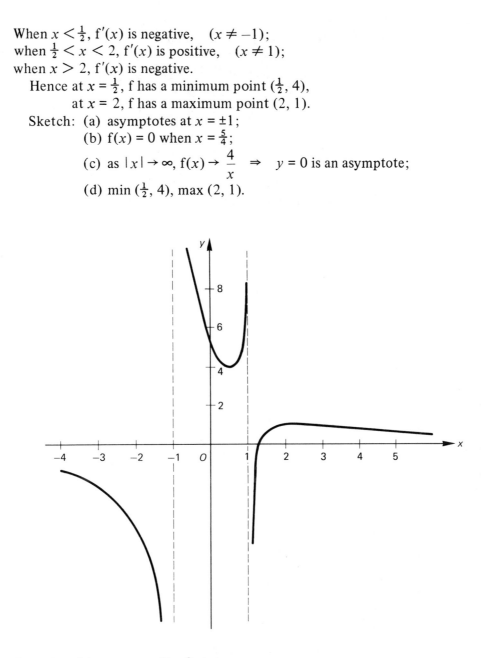

Domain of f is ($x : x \in \mathbb{R}, x^2 \neq 1$).
Range is f(x) $\leqslant 1$ or f(x) $\geqslant 4$.

15.7 $V = \pi r^2 h.$ (1)
Surface area = $A = 2\pi r^2 + 2\pi rh.$ (2)

  From equation 1, $h = \dfrac{V}{\pi r^2}$; in equation 2, $A = 2\pi r^2 + \dfrac{2\pi r V}{\pi r^2} = 2\pi r^2 + \dfrac{2V}{r}$.

$\dfrac{\mathrm{d}A}{\mathrm{d}r} = 4\pi r - \dfrac{2V}{r^2} = 0$ when $4\pi r = \dfrac{2V}{r^2}$ $\Rightarrow$ $r^3 = \dfrac{V}{2\pi}$.

$\dfrac{\mathrm{d}^2 A}{\mathrm{d}r^2} = 4\pi + \dfrac{4V}{r^3} > 0$ for all $r.$

$h^3 = \dfrac{4V}{\pi}, \dfrac{r^3}{h^3} = \dfrac{1}{8}, r : h = 1 : 2.$

15.8 $\dfrac{dx}{d\theta} = 2(1 - \cos\theta),\ \dfrac{dy}{d\theta} = 2\sin\theta,$

$\dfrac{dy}{dx} = \dfrac{2\sin\theta}{2(1 - \cos\theta)} = \dfrac{2\sin\theta/2\cos\theta/2}{2\sin^2\theta/2} = \cot\theta/2.$

At A, $\theta = \dfrac{\pi}{2}$, $x = \pi - 2$, $y = 2$, $\dfrac{dy}{dx} = 1$.

At B, $\theta = \dfrac{3\pi}{2}$, $x = 3\pi + 2$, $y = 2$, $\dfrac{dy}{dx} = -1$.

Equation of tangent at A is $y = x - \pi + 4$.
Equation of tangent at B is $y = -x + 3\pi + 4$.

16 Integration—1

Integration as the inverse of differentiation. Integration of standard functions. Simple techniques of integration including partial fractions. The evaluation of definite integrals.

16.1 Fact Sheet

(a) Standard Results (Arbitrary Constants Omitted)

$f(x)$	$\int f(x)\,dx$
$x^n \ (n \neq -1)$	$\dfrac{x^{n+1}}{n+1}$
$\dfrac{1}{x}$	$\ln|x|$
e^x	e^x
$\sin x$	$-\cos x$
$\cos x$	$\sin x$
$\tan x$	$\ln|\sec x|$ or $-\ln|\cos x|$
$\cot x$	$-\ln|\text{cosec } x|$ or $\ln|\sin x|$
$\sec^2 x$	$\tan x$
$\text{cosec}^2 x$	$-\cot x$
$\sec x$	$\ln|\sec x + \tan x|$ or $\ln\left|\tan\left(\dfrac{x}{2} + \dfrac{\pi}{4}\right)\right|$
$\text{cosec } x$	$-\ln|\text{cosec } x + \cot x|$ or $\ln\left|\cot\dfrac{x}{2}\right|$
$\sec x \tan x$	$\sec x$
$\text{cosec } x \cot x$	$-\text{cosec } x$
$\dfrac{1}{1+x^2}$	$\tan^{-1} x$
$\dfrac{1}{\sqrt{(1-x^2)}}$	$\sin^{-1} x$

(b) Integration as the Inverse of Differentiation

$$\int f'(x)\,dx = f(x).$$

$$\int \{f(x)\}^n \, f'(x)\,dx = \frac{\{f(x)\}^{n+1}}{n+1}.$$

$$\int \frac{f'(x)}{f(x)} \, dx = \ln |f(x)|.$$

$$\int \{f'(x)\} \{e^{f(x)}\} \, dx = e^{f(x)}.$$

(c) Use of Partial Fractions

$$\int \frac{f(x)}{g(x)} \, dx$$ where $f(x)$ and $g(x)$ are polynomials and $g(x)$ will factorize.

If $f(x)$ is of the same degree as $g(x)$ or a higher one then long division must be carried out before changing to partial fractions.

(d) General Techniques for Integrating Trigonometric Functions

(i) $\sin^n x$ (and $\cos^n x$) where n is an even integer: use the $\cos 2x$ formulae.
(ii) $\sin^n x$ (and $\cos^n x$) where n is an odd integer: write $\sin^n x = \sin^{n-1} x \sin x$ and then change the $\sin^{n-1} x$ into terms in $\cos x$ using $\sin^2 x \equiv 1 - \cos^2 x$.
(iii) $\tan^n x$ where n is an integer and $n > 1$: use $\tan^2 x \equiv \sec^2 x - 1$.
(iv) $\sin ax \cos bx$ and similar terms: use the product formulae.

(e) Indefinite Integrals

If $\dfrac{d}{dx} \{F(x)\} = f(x)$ then $\displaystyle\int f(x) \, dx = F(x) + $ constant.

(f) Definite Integrals

$$\int_a^b f(x) \, dx = \left[F(x) \right]_a^b = F(b) - F(a).$$

16.2   Worked Examples

16.1

(a) Evaluate (i) $\displaystyle\int_1^2 \frac{dx}{(2x+1)^2}$ and (ii) $\displaystyle\int_1^2 \frac{x}{(2x+1)^2} \, dx.$

(b) Evaluate $\displaystyle\int_0^{\pi/3} \sin 2x \cos x \, dx.$

● (a) (i) $\displaystyle\int_1^2 \frac{dx}{(2x+1)^2} = \frac{1}{2} \int_1^2 \frac{2}{(2x+1)^2} \, dx$

$$= \frac{1}{2} \left[\frac{-1}{(2x+1)} \right]_1^2 = \frac{-1}{2(5)} + \frac{1}{2(3)} = \frac{1}{15}.$$

(ii) $\displaystyle\int_1^2 \frac{x}{(2x+1)^2}\,dx = \frac{1}{2}\int_1^2 \left\{\frac{2x+1}{(2x+1)^2} - \frac{1}{(2x+1)^2}\right\}\,dx$

$$= \frac{1}{2}\int_1^2 \left\{\frac{1}{2x+1} - \frac{1}{(2x+1)^2}\right\}\,dx$$

$$= \frac{1}{4}\int_1^2 \left\{\frac{2}{2x+1} - \frac{2}{(2x+1)^2}\right\}\,dx$$

$$= \frac{1}{4}\left[\ln|2x+1| + \frac{1}{2x+1}\right]_1^2$$

$$= \tfrac{1}{4}(\ln 5 + \tfrac{1}{5} - \ln 3 - \tfrac{1}{3})$$

$$= \tfrac{1}{4}(\ln \tfrac{5}{3} - \tfrac{2}{15}) = 0.094.$$

(b) $\displaystyle\int_0^{\pi/3} \sin 2x \cos x\,dx$

Products $\sin ax \cos bx$ and similar usually require the product formulae.

$$\sin 2x \cos x = \tfrac{1}{2}(\sin 3x + \sin x).$$

$$\int_0^{\pi/3} \sin 2x \cos x\,dx = \tfrac{1}{2}\int_0^{\pi/3} (\sin 3x + \sin x)\,dx$$

$$= \frac{1}{2}\left[\frac{-\cos 3x}{3} - \cos x\right]_0^{\pi/3}$$

$$= \frac{1}{2}\left(\frac{-\cos \pi}{3} - \cos \frac{\pi}{3} + \frac{\cos 0}{3} + \cos 0\right)$$

$$= \tfrac{1}{2}(\tfrac{1}{3} - \tfrac{1}{2} + \tfrac{1}{3} + 1)$$

$$= \tfrac{7}{12}.$$

16.2

(a) Show that $\displaystyle\int_4^6 \frac{x^2 + x - 11}{(x-2)^2\,(x-3)}\,dx = \ln a + \frac{b}{c}$

a, b and c are integers to be determined.

(b) Find $\displaystyle\int \sin x \cos x\,dx$ by

(i) considering the result of differentiating $\sin^2 x$,
(ii) considering the result of differentiating $\cos 2x$.
Why are the results apparently different?

● (a) By partial fractions:

$$\frac{x^2 + x - 11}{(x-2)^2\,(x-3)} \equiv \frac{A}{x-2} + \frac{B}{(x-2)^2} + \frac{C}{x-3}$$

$$\equiv \frac{A(x-2)(x-3) + B(x-3) + C(x-2)^2}{(x-2)^2\,(x-3)}.$$

Equating numerators:

$$x^2 + x - 11 \equiv A\,(x-2)\,(x-3) + B\,(x-3) + C\,(x-2)^2.$$

Let $x = 3$; then $9 + 3 - 11 = C(1) \quad \Rightarrow \quad C = 1$

$x = 2$; then $4 + 2 - 11 = B(-1) \quad \Rightarrow \quad B = 5$

$x = 0$; then $-11 = 6A - 3B + 4C = 6A - 15 + 4 \quad \Rightarrow \quad A = 0.$

Hence
$$\int_4^6 \frac{x^2 + x - 11}{(x-2)^2\,(x-3)}\,dx = \int_4^6 \left\{ \frac{5}{(x-2)^2} + \frac{1}{x-3} \right\}\,dx$$

$$= \left[\frac{-5}{(x-2)} + \ln|x-3| \right]_4^6$$

$$= -\tfrac{5}{4} + \ln 3 + \tfrac{5}{2} - \ln 1$$

$$= \ln 3 + \tfrac{5}{4}.$$

$\Rightarrow \quad a = 3,\ b = 5 \text{ and } c = 4.$

(b) $\displaystyle\int \sin x \cos x \, dx.$

$$\frac{d}{dx}(\sin^2 x) = 2 \sin x \cos x \quad \Rightarrow \quad \int \sin x \cos x \, dx = \tfrac{1}{2}\sin^2 x + C_1.$$

$$\frac{d}{dx}(\cos 2x) = -2 \sin 2x = -4 \sin x \cos x$$

$$\Rightarrow \quad \int \sin x \cos x \, dx = -\tfrac{1}{4}\cos 2x + C_2.$$

The results appear to be different but the difference is in the constant:

$$\cos 2x = 1 - 2\sin^2 x \quad \Rightarrow \quad -\tfrac{1}{4}\cos 2x = -\tfrac{1}{4} + \tfrac{1}{2}\sin^2 x$$

$$= \tfrac{1}{2}\sin^2 x + \text{constant}.$$

16.3 Find (a) $\displaystyle\int \sin^2 x \, dx$, (b) $\displaystyle\int \cos^3 x \, dx$ and (c) $\displaystyle\int \sin^2 x \cos^3 x \, dx.$

● (a)
$$\sin^2 x = \frac{1 - \cos 2x}{2}$$

$$\Rightarrow \quad \int \sin^2 x \, dx = \tfrac{1}{2} \int (1 - \cos 2x) \, dx$$

$$= \tfrac{1}{2}\left(x - \frac{\sin 2x}{2} \right) + C$$

$$= \tfrac{1}{4}(2x - \sin 2x) + C.$$

(b)
$$\cos^3 x = \cos^2 x \cos x = (1 - \sin^2 x) \cos x$$

$$\Rightarrow \quad \int \cos^3 x \, dx = \int (\cos x - \sin^2 x \cos x) \, dx$$

$$= \sin x - \frac{\sin^3 x}{3} + C.$$

(c)
$$\sin^2 x \cos^3 x = \sin^2 x\,(1 - \sin^2 x)\cos x$$

$$= \sin^2 x \cos x - \sin^4 x \cos x$$

$$\Rightarrow \int \sin^2 x \cos^3 x\, dx = \int (\sin^2 x \cos x - \sin^4 x \cos x)\, dx$$

$$= \frac{\sin^3 x}{3} - \frac{\sin^5 x}{5} + C.$$

16.4

(a) Show that $\displaystyle\int_0^{\pi/4} (1 + \tan x)^2\, dx = 1 + \ln 2.$

(b) Evaluate $\displaystyle\int_0^{\pi/2} \frac{1}{\sin x + \cos x}\, dx$, giving the answer in exact form.

● (a) $(1 + \tan x)^2 = 1 + \tan^2 x + 2 \tan x = \sec^2 x + 2 \tan x.$

$$\int_0^{\pi/4} (1 + \tan x)^2\, dx = \int_0^{\pi/4} (\sec^2 x + 2 \tan x)\, dx$$

$$= \left[\tan x + 2 \ln |\sec x| \right]_0^{\pi/4}$$

$$= \tan \frac{\pi}{4} + 2 \ln \left| \sec \frac{\pi}{4} \right| - \tan 0 - 2 \ln |\sec 0|$$

$$= 1 + 2 \ln \sqrt{2} = 1 + \ln 2.$$

(b) $a \cos x + b \sin x$ or $a \sin x + b \cos x$ can be expressed as $r \cos (x - \alpha)$ (see Chapter 9).

$$\frac{1}{\sin x + \cos x} = \frac{1}{\sqrt{2} \cos (x - \pi/4)} = \frac{1}{\sqrt{2}} \sec \left(x - \frac{\pi}{4} \right).$$

$$\int_0^{\pi/2} \frac{1}{\sin x + \cos x}\, dx = \frac{1}{\sqrt{2}} \int_0^{\pi/2} \sec \left(x - \frac{\pi}{4} \right)\, dx$$

$$= \frac{1}{\sqrt{2}} \left[\ln \left| \sec \left(x - \frac{\pi}{4} \right) + \tan \left(x - \frac{\pi}{4} \right) \right| \right]_0^{\pi/2}$$

$$= \frac{1}{\sqrt{2}} \left[\ln \left| \sec \frac{\pi}{4} + \tan \frac{\pi}{4} \right| \right.$$

$$\left. - \ln \left| \sec \left(-\frac{\pi}{4} \right) + \tan \left(-\frac{\pi}{4} \right) \right| \right]$$

$$= \frac{1}{\sqrt{2}} \left\{ \ln (\sqrt{2} + 1) - \ln (\sqrt{2} - 1) \right\}$$

$$= \frac{1}{\sqrt{2}} \ln \frac{\sqrt{2} + 1}{\sqrt{2} - 1}.$$

But $\displaystyle\frac{1}{\sqrt{2} - 1} = \frac{\sqrt{2} + 1}{(\sqrt{2} - 1)(\sqrt{2} + 1)} = \sqrt{2} + 1.$

Hence $\displaystyle\int_0^{\pi/2} \frac{1}{\sin x + \cos x}\, dx = \frac{1}{\sqrt{2}} \ln (\sqrt{2} + 1)^2 = (\sqrt{2}) \ln (\sqrt{2} + 1).$

16.5 Evaluate $\displaystyle\int_0^{\pi/6} \sin x \sec^2 x \, dx.$

● $\sin x \sec^2 x = \dfrac{\sin x}{\cos^2 x} = \sec x \tan x.$

$$\int_0^{\pi/6} \sin x \sec^2 x \, dx = \int_0^{\pi/6} \sec x \tan x \, dx$$

$$= \Big[\sec x \Big]_0^{\pi/6}$$

$$= \frac{2}{\sqrt{3}} - 1.$$

Alternatively,

$$\int_0^{\pi/6} \sin x \sec^2 x \, dx = \int_0^{\pi/6} \cos^{-2} x \sin x \, dx$$

$$= -\int_0^{\pi/6} \cos^{-2} x \, d(\cos x)$$

$$= \Big[(\cos x)^{-1} \Big]_0^{\pi/6}$$

$$= \frac{2}{\sqrt{3}} - 1.$$

16.6 Find $\displaystyle\int \tan^3 x \, dx.$

● $\tan^3 x = \tan^2 x \tan x = (\sec^2 x - 1) \tan x$
$= \sec^2 x \tan x - \tan x.$

$$\int \tan^3 x \, dx = \int (\sec^2 x \tan x - \tan x) \, dx$$

$$= \tfrac{1}{2} \tan^2 x + \ln |\cos x| + C.$$

16.7

(a) Find (i) $\displaystyle\int (\cos 3x + \sin 3x)^2 \, dx,$ (ii) $\displaystyle\int (\cos 3x + \sin 2x)^2 \, dx.$

(b) Evaluate $\displaystyle\int_2^3 x \, (x - 2)^{12} \, dx.$

● (a) (i) $(\cos 3x + \sin 3x)^2 = \cos^2 3x + \sin^2 3x + 2 \sin 3x \cos 3x$
$= 1 + \sin 6x.$

$$\int (\cos 3x + \sin 3x)^2 \, dx = \int (1 + \sin 6x) \, dx$$

$$= x - \frac{\cos 6x}{6} + C.$$

(ii) $(\cos 3x + \sin 2x)^2 = \cos^2 3x + \sin^2 2x + 2 \cos 3x \sin 2x$

$$= \frac{1 + \cos 6x}{2} + \frac{1 - \cos 4x}{2} + \sin 5x - \sin x.$$

$$\int (\cos 3x + \sin 2x)^2 \, dx = \int \left(1 + \frac{\cos 6x}{2} - \frac{\cos 4x}{2} + \sin 5x - \sin x \right) dx$$

$$= x + \frac{\sin 6x}{12} - \frac{\sin 4x}{8} - \frac{\cos 5x}{5} + \cos x + C.$$

(b) $\int_2^3 x \, (x - 2)^{12} \, dx$. Write x in the form $[(x - 2) + 2]$.

Then
$$x \, (x - 2)^{12} = (x - 2) \, (x - 2)^{12} + 2 \, (x - 2)^{12}$$
$$= (x - 2)^{13} + 2 \, (x - 2)^{12}.$$

$$\int_2^3 x \, (x - 2)^{12} \, dx = \int_2^3 \{ (x - 2)^{13} + 2 \, (x - 2)^{12} \} \, dx$$

$$= \left[\frac{(x - 2)^{14}}{14} + \frac{2 \, (x - 2)^{13}}{13} \right]_2^3$$

$$= \tfrac{1}{14} + \tfrac{2}{13} = \tfrac{41}{182}.$$

Alternatively, use a substitution $u = x - 2$ (see Chapter 17).

16.8

(a) Evaluate (i) $\int_{-1}^2 \frac{x}{\sqrt{(x + 2)}} \, dx$,   (ii) $\int_1^8 \left(\sqrt[3]{x} + \frac{1}{2 \sqrt[3]{x}} \right) dx$.

(b) Evaluate $\int_0^\pi (\sin^2 x + \cos^2 x)^2 \, dx$.

● (a) (i) $\int_{-1}^2 \frac{x}{\sqrt{(x + 2)}} \, dx.$

$$\frac{x}{\sqrt{(x + 2)}} = \frac{x + 2 - 2}{\sqrt{(x + 2)}} = (x + 2)^{1/2} - 2 \, (x + 2)^{-1/2}.$$

$$\int_{-1}^2 \frac{x}{\sqrt{(x + 2)}} \, dx = \int_{-1}^2 \{ (x + 2)^{1/2} - 2 \, (x + 2)^{-1/2} \} \, dx$$

$$= \left[\tfrac{2}{3} (x + 2)^{3/2} - (2) \{ 2 \, (x + 2)^{1/2} \} \right]_{-1}^2$$

$$= \{ \tfrac{2}{3} (4)^{3/2} - 4 \, (4)^{1/2} \} - \{ \tfrac{2}{3} (1)^{3/2} - 4 \, (1)^{1/2} \} = \tfrac{2}{3}.$$

Alternatively, use a substitution.

(ii) $\int_1^8 \left(\sqrt[3]{x} + \frac{1}{2 \sqrt[3]{x}} \right) dx = \int_1^8 \{ (x)^{1/3} + \tfrac{1}{2} (x)^{-1/3} \} \, dx$

$$= \left[\tfrac{3}{4}(x)^{4/3} + (\tfrac{1}{2})\,(\tfrac{3}{2})\,(x)^{2/3} \right]_1^8$$

$$= \{ \tfrac{3}{4}(8)^{4/3} + \tfrac{3}{4}(8)^{2/3} \} - \{ \tfrac{3}{4} + \tfrac{3}{4} \} = \tfrac{27}{2}.$$

(b) **Did you fall into the trap?**

$$\int_0^\pi (\sin^2 x + \cos^2 x)^2 \; dx = \int_0^\pi (1)\; dx = \left[x \right]_0^\pi = \pi.$$

16.9 Evaluate   (a) $\displaystyle\int_{-1}^1 \frac{x}{(x+3)}\; dx$;   (b) $\displaystyle\int_0^\pi \cos^4 x \; dx$;

(c) $\displaystyle\int_{\sqrt{2}}^{\sqrt{3}} \frac{x^3}{x^2 - 1}\; dx$;   (d) $\displaystyle\int_{\pi/4}^{\pi/2} \frac{1}{\sin^2 x}\; dx$.

● (a) $$\frac{x}{x+3} = \frac{x+3-3}{x+3} = 1 - \frac{3}{x+3}.$$

$$\int_{-1}^1 \frac{x}{x+3}\; dx = \int_{-1}^1 \left(1 - \frac{3}{x+3} \right) dx.$$

$$= \left[x - 3 \ln |x + 3| \right]_{-1}^1$$

$$= (1 - 3 \ln 4) - (-1 - 3 \ln 2)$$

$$= 2 - 3 \ln 2 = -0.079.$$

(b) $$\cos^4 x = \frac{(1 + \cos 2x)^2}{4} = \frac{1 + 2 \cos 2x + \cos^2 2x}{4}$$

$$= \tfrac{1}{4} + \tfrac{1}{2} \cos 2x + \frac{1 + \cos 4x}{8}.$$

$$\int_0^\pi \cos^4 x \; dx = \int_0^\pi (\tfrac{3}{8} + \tfrac{1}{2} \cos 2x + \tfrac{1}{8} \cos 4x)\; dx$$

$$= \left[\frac{3x}{8} + \frac{\sin 2x}{4} + \frac{\sin 4x}{32} \right]_0^\pi = \frac{3}{8}\, \pi.$$

(c) $$\frac{x^3}{x^2 - 1} = x + \frac{x}{x^2 - 1} \quad \text{by long division.}$$

$$\int_{\sqrt{2}}^{\sqrt{3}} \frac{x^3}{x^2 - 1}\; dx = \int_{\sqrt{2}}^{\sqrt{3}} \left(x + \frac{x}{x^2 - 1} \right) dx$$

$$= \left[\frac{x^2}{2} + \tfrac{1}{2} \ln |x^2 - 1| \right]_{\sqrt{2}}^{\sqrt{3}}$$

$$= (\tfrac{3}{2} + \tfrac{1}{2} \ln (2)) - (\tfrac{2}{2} + \tfrac{1}{2} \ln (1))$$

$$= \tfrac{1}{2} + \tfrac{1}{2} \ln 2 = 0.847.$$

(d) $\displaystyle\int_{\pi/4}^{\pi/2} \frac{1}{\sin^2 x}\ dx = \int_{\pi/4}^{\pi/2} \operatorname{cosec}^2 x\ dx$

$$= \left[-\cot x\right]_{\pi/4}^{\pi/2} = (0) - (-1) = 1.$$

16.10 Given that $f(x) \equiv \dfrac{3x + 2}{(2x - 1)(x + 3)}$, express $f(x)$ in partial fractions.

Sketch the curve $y = f(x)$, showing the asymptotes and the points of intersection of the curve with the axes.

Evaluate $\displaystyle\int_1^5 f(x)\ dx$ and shade on your sketch the region whose area is equal to this integral.

(L)

● $\dfrac{3x + 2}{(2x - 1)(x + 3)} \equiv \dfrac{A}{(2x - 1)} + \dfrac{B}{(x + 3)}$

$$\equiv \dfrac{A(x + 3) + B(2x - 1)}{(2x - 1)(x + 3)}.$$

Comparing the numerators: let $x = -3$; $\quad -7 = B(-7) \quad \Rightarrow \quad B = 1$;

let $x = \tfrac{1}{2}$; $\quad \tfrac{7}{2} = A\left(\tfrac{7}{2}\right) \quad \Rightarrow \quad A = 1.$

$$f(x) \equiv \dfrac{1}{2x - 1} + \dfrac{1}{x + 3}.$$

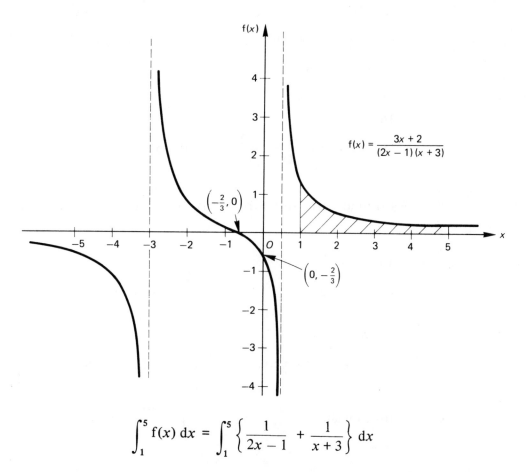

$$\int_1^5 f(x)\ dx = \int_1^5 \left\{\frac{1}{2x - 1} + \frac{1}{x + 3}\right\} dx$$

185

$$= \left[\tfrac{1}{2} \ln |2x - 1| + \ln |x + 3| \right]_1^5$$

$$= (\tfrac{1}{2} \ln 9 + \ln 8) - (\tfrac{1}{2} \ln 1 + \ln 4)$$

$$= \ln 3 + \ln 8 - \ln 4$$

$$= \ln 6$$

$$= 1.792.$$

16.3 Exercises

16.1 For which of the following is it true that

$$\int_{-1}^{1} f(x)\, dx = 2 \int_{0}^{1} f(x)\, dx?$$

(i) $f(x) = \sin x$,   (ii) $f(x) = \cos x$,   (iii) $f(x) = x^2$,   (iv) $f(x) = x^3$.

A, (i) only;   B, (ii) and (iv) only;
C, (iii) and (iv) only;   D, (ii) and (iii) only;
E, some other combination.

16.2 Given that K is a constant of integration, then,

for $x > 0, \displaystyle\int (1 + x^{1/2} - x^{-3/2})\, dx$ equals

A, $\dfrac{2x^{3/2}}{3} + 2x^{-1/2} + K$;   B, $x + \tfrac{2}{3}x^{3/2} + 2x^{-1/2} + K$;

C, $x + \tfrac{3}{2}x^{3/2} + \tfrac{1}{2}x^{-1/2} + K$;   D, $x + \tfrac{2}{3}x^{3/2} - 2x^{-1/2} + K$;

E, none of these.

16.3

(a) Evaluate $\displaystyle\int_{1}^{2} \dfrac{2}{(x - 4)(x - 6)}\, dx$ to three significant figures.

(b) Find   (i) $\displaystyle\int \sin^3 x\, dx$,   (ii) $\displaystyle\int \tan^2 x\, dx$,   (iii) $\displaystyle\int \cot^3 x\, dx$.

16.4

(a) Evaluate $\displaystyle\int_{0}^{\pi/2} \cos x\, dx.$

Write down

$$\int_{-\pi/2}^{\pi/2} \cos x\, dx; \qquad \int_{0}^{\pi} \cos x\, dx; \qquad \int_{0}^{\pi} \sin x\, dx; \qquad \int_{0}^{\pi/2} \sin 2x\, dx.$$

(b) Evaluate $\displaystyle\int_{2}^{3} \dfrac{x^2}{x - 1}\, dx.$

16.5

(a) Evaluate
$$\int_0^{\pi/2} \sin^2 x \, dx, \qquad \int_0^{\pi/2} \cos^2 x \, dx,$$

$$\int_0^{\pi/2} \cos^2 2x \, dx, \qquad \int_0^{\pi/2} \cos^2 ax \, dx,$$

where a is an integer.

(b) Find
$$\int \frac{x+2}{(x-1)^2 \, (x+1)} \, dx.$$

16.6 Given that $f(x) \equiv \dfrac{x^2}{(x+1)^2 \, (x-1)}$, express $f(x)$ in partial fractions. Hence:

(a) Find $\int f(x) \, dx$.

(b) Find the first three terms $ax^2 + bx^3 + cx^4$ when $f(x)$ is expanded in a series of ascending powers of x, stating the values of x for which the expansion is valid.

(c) Evaluate $\displaystyle\int_0^{0.2} f(x) \, dx$ and $\displaystyle\int_0^{0.2} (ax^2 + bx^3 + cx^4) \, dx.$

16.7 Evaluate

(a) $\displaystyle\int_1^4 \frac{(\sqrt{x}+1)^3}{5\sqrt{x}} \, dx,$ (b) $\displaystyle\int_0^{\pi/2} \sin^3 x \cos^4 x \, dx,$

(c) $\displaystyle\int_0^{\sqrt{3}/2} \frac{2x}{\sqrt{(1-x^2)}} \, dx,$ (d) $\displaystyle\int_{-1}^1 \frac{1+e^{2x}}{e^x} \, dx.$

16.8 Evaluate

(a) $\displaystyle\int_0^{\pi/12} \frac{1}{\cos^2 3x} \, dx,$ (b) $\displaystyle\int_0^1 \frac{x^3}{x^2+1} \, dx.$

16.9 Given that $f(x) \equiv \dfrac{a}{(x-3)\,(x+1)}$, express $f(x)$ as partial fractions.

If $\displaystyle\int_0^1 f(x) \, dx = \ln\left(\tfrac{1}{9}\right),$ find a.

Using this value of a, find the value of b for which

$$\int_0^b f(x) \, dx = 2 \ln \tfrac{1}{9}.$$

16.4   Outline Solutions to Exercises

16.1
$$\int_{-1}^{1} f(x)\,dx = 2 \int_{0}^{1} f(x)\,dx$$

if $f(x)$ is an even function. $\cos x$ and x^2 are even functions.   <u>**Answer D**</u>

16.2
$$\int (1 + x^{1/2} - x^{-3/2})\,dx = x + \tfrac{2}{3}x^{3/2} + 2x^{-1/2} + K.$$ <u>**Answer B**</u>

16.3

(a)
$$\int_{1}^{2} \frac{2}{(x-4)(x-6)}\,dx = \int_{1}^{2}\left(\frac{-1}{x-4} + \frac{1}{x-6}\right)dx$$
$$= \left[-\ln|x-4| + \ln|x-6|\right]_{1}^{2} = \ln \tfrac{6}{5} = 0.182.$$

(b) (i) $\displaystyle\int \sin^3 x\,dx = \int (1 - \cos^2 x)\sin x\,dx = -\cos x + \frac{\cos^3 x}{3} + C.$

(ii) $\displaystyle\int \tan^2 x\,dx = \int (\sec^2 x - 1)\,dx = \tan x - x + C.$

(iii) $\displaystyle\int \cot^3 x\,dx = \int \cot x\,(\operatorname{cosec}^2 x - 1)\,dx$
$$= -\tfrac{1}{2}\cot^2 x + \ln|\operatorname{cosec} x| + C.$$

16.4

(a) $\displaystyle\int_{0}^{\pi/2} \cos x\,dx = \left[\sin x\right]_{0}^{\pi/2} = 1.$

2; 0; 2; 1. (Think of the graphs.)

(b) $\displaystyle\int_{2}^{3}\left(x + 1 + \frac{1}{x-1}\right)dx = \left[\frac{x^2}{2} + x + \ln|x-1|\right]_{2}^{3} = \tfrac{7}{2} + \ln 2 = 4.193.$

16.5

(a) $\displaystyle\int_{0}^{\pi/2} \sin^2 x\,dx = \frac{1}{2}\int_{0}^{\pi/2}(1 - \cos 2x)\,dx = \frac{1}{2}\left[x - \frac{\sin 2x}{2}\right]_{0}^{\pi/2} = \frac{\pi}{4}.$

$\displaystyle\int_{0}^{\pi/2} \cos^2 x\,dx = \frac{1}{2}\int_{0}^{\pi/2}(1 + \cos 2x)\,dx = \frac{1}{2}\left[x + \frac{\sin 2x}{2}\right]_{0}^{\pi/2} = \frac{\pi}{4}.$

$\displaystyle\int_{0}^{\pi/2} \cos^2 2x\,dx = \frac{1}{2}\int_{0}^{\pi/2}(1 + \cos 4x)\,dx = \frac{\pi}{4}.$

$\displaystyle\int_{0}^{\pi/2} \cos^2 ax\,dx = \frac{1}{2}\int_{0}^{\pi/2}(1 + \cos 2ax)\,dx = \frac{\pi}{4}.$

(b) $\displaystyle\frac{1}{4}\int\left\{\frac{1}{x+1} + \frac{-1}{x-1} + \frac{6}{(x-1)^2}\right\}dx = \frac{1}{4}\ln\left|\frac{x+1}{x-1}\right| - \frac{3}{2(x-1)} + C.$

16.6 $f(x) = \dfrac{1}{4}\left\{\dfrac{1}{x-1} + \dfrac{3}{x+1} - \dfrac{2}{(x+1)^2}\right\}.$

(a) $\dfrac{1}{2(x+1)} + \dfrac{3}{4}\ln|x+1| + \dfrac{1}{4}\ln|x-1| + C$

(b) $-x^2(1 - 2x + 3x^2 - \ldots)(1 + x + x^2 + \ldots)$

  $= -x^2 + x^3 - 2x^4 + \ldots$ (valid when $-1 < x < 1$).

(c) $\displaystyle\int_0^{0.2} f(x)\,dx = \dfrac{1}{2(1.2)} + \dfrac{3}{4}\ln(1.2) + \dfrac{1}{4}\ln(0.8) - \dfrac{1}{2} + 0 + 0$

  $= -0.002\,38.$

  $\displaystyle\int_0^{0.2}(-x^2 + x^3 - 2x^4)\,dx = \left[\dfrac{-x^3}{3} + \dfrac{x^4}{4} - \dfrac{2x^5}{5}\right]_0^{0.2} = -0.002\,39.$

16.7

(a) $\dfrac{1}{5}\displaystyle\int_1^4(x + 3x^{1/2} + 3 + x^{-1/2})\,dx = \tfrac{13}{2}.$

(b) $\displaystyle\int_0^{\pi/2}(\cos^4 x \sin x - \cos^6 x \sin x)\,dx = \left[-\dfrac{\cos^5 x}{5} + \dfrac{\cos^7 x}{7}\right]_0^{\pi/2} = \tfrac{2}{35}.$

(c) $\left[-2(1-x^2)^{1/2}\right]_0^{\sqrt{3}/2} = -2(\tfrac{1}{4})^{1/2} + 2(1)^{1/2} = 1.$

(d) $\displaystyle\int_{-1}^1(e^{-x} + e^x)\,dx = \left[-e^{-x} + e^x\right]_{-1}^1$

  $= (-e^{-1} + e^1) - (-e^1 + e^{-1}) = 4.70.$

16.8

(a) $\displaystyle\int_0^{\pi/12}\sec^2 3x\,dx = \left[\tfrac{1}{3}\tan 3x\right]_0^{\pi/12} = \tfrac{1}{3}.$

(b) $\displaystyle\int_0^1\left(x - \dfrac{x}{x^2+1}\right)dx = \left[\dfrac{x^2}{2} - \dfrac{1}{2}\ln|x^2+1|\right]_0^1 = \dfrac{1}{2} - \dfrac{1}{2}\ln 2 = 0.153.$

16.9 $\dfrac{1}{4}\left(\dfrac{a}{x-3} - \dfrac{a}{x+1}\right).$

  $\displaystyle\int_0^1 f(x)\,dx = \dfrac{a}{4}\left[\ln|x-3| - \ln|x+1|\right]_0^1 = -\dfrac{a}{4}\ln 3$

  $\Rightarrow\ a = 8.$

  $\displaystyle\int_0^b f(x)\,dx = 2\left[\ln|b-3| - \ln|b+1| - \ln|-3| + \ln 1\right]$

  $= 2\ln\dfrac{(3-b)}{3(b+1)}$ $(-1 < b < 3).$

  $\Rightarrow\ \dfrac{(3-b)}{3(b+1)} = \dfrac{1}{9}$ $\Rightarrow\ b = 2.$

17 Integration—2

Integration by substitution and by parts.

17.1 Fact Sheet

(a) Integration by Substitution

$$\int f(x)\,dx = \int f(x)\,\frac{dx}{du}\,du \text{ where } f(x)\,\frac{dx}{du} \text{ is expressed in terms of } u.$$

- With indefinite integrals, answers must be expressed in terms of the original variable.
- With definite integrals remember to change the limits of integration to the corresponding values of the new variable.

(b) Special Substitutions

(i) $\displaystyle\int \frac{1}{\sqrt{(a^2 - b^2 x^2)}}\,dx$ or $\displaystyle\int \sqrt{(a^2 - b^2 x^2)}\,dx.$

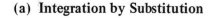

Let $bx = a\sin\theta$; then $b\,\dfrac{dx}{d\theta} = a\cos\theta \Rightarrow dx = \dfrac{a}{b}\cos\theta\,d\theta$

and $a^2 - b^2 x^2 = a^2\cos^2\theta.$

(ii) $\displaystyle\int \frac{1}{a^2 + b^2 x^2}\,dx.$

Let $bx = a\tan\theta$; then $b\,\dfrac{dx}{d\theta} = a\sec^2\theta \Rightarrow dx = \dfrac{a}{b}\sec^2\theta\,d\theta$

and $a^2 + b^2 x^2 = a^2\sec^2\theta.$

(iii) $\displaystyle\int \frac{1}{a\sin\theta + b\cos\theta + c}\,d\theta.$

Let $t = \tan\dfrac{\theta}{2}$; then $\dfrac{dt}{d\theta} = \dfrac{1}{2}\sec^2\dfrac{\theta}{2} = \dfrac{(1 + t^2)}{2} \Rightarrow d\theta = \dfrac{2\,dt}{1 + t^2}$

and $\sin\theta = \dfrac{2t}{1 + t^2}$, $\cos\theta = \dfrac{1 - t^2}{1 + t^2}.$

(iv) $\displaystyle\int \frac{1}{a\cos^2\theta + b\sin^2\theta + c}\,d\theta.$

Let $t = \tan\theta$; then $\dfrac{dt}{d\theta} = \sec^2\theta = (1 + t^2) \Rightarrow d\theta = \dfrac{dt}{1 + t^2}$

$$\text{and } \sin^2 \theta = \frac{t^2}{1 + t^2}, \quad \cos^2 \theta = \frac{1}{1 + t^2}.$$

(v) $\displaystyle\int \sin^n \theta \cos \theta \; d\theta$: let $s = \sin \theta$; then $ds = \cos \theta \; d\theta$.

$\displaystyle\int \cos^n \theta \sin \theta \; d\theta$: let $c = \cos \theta$; then $dc = -\sin \theta \; d\theta$.

$\displaystyle\int \tan^n \theta \sec^2 \theta \; d\theta$: let $t = \tan \theta$; then $dt = \sec^2 \theta \; d\theta$.

$\displaystyle\int \sec^n \theta \tan \theta \; d\theta$: let $s = \sec \theta$; then $ds = \sec \theta \tan \theta \; d\theta$.

(vi) $\displaystyle\int \frac{1}{x^2 + 2bx + c} \; dx$, where $x^2 + 2bx + c$ will not readily factorize,

$$x^2 + 2bx + c = (x + b)^2 + (c - b^2).$$

If $(c - b^2) > 0$, the denominator is the sum of two squares. Use a tangent substitution for $(x + b)$ as in (ii).

If $(c - b^2) < 0$ use partial fractions as in Chapter 16.

(vii) $\displaystyle\int \frac{1}{\sqrt{(a + 2bx - x^2)}} \; dx$.

Change $a + 2bx - x^2$ to $(a + b^2) - (x - b)^2$ and use a sine substitution for $(x - b)$ as in (i).

(viii) $\displaystyle\int x\sqrt{(ax + b)} \; dx \qquad \text{or} \qquad \int \frac{f(x)}{\sqrt{(ax + b)}} \; dx$.

Let $u = \sqrt{(ax + b)}$; square both sides: $u^2 = ax + b$;

$$\text{then } 2u \frac{du}{dx} = a \quad \Rightarrow \quad dx = \frac{2u}{a} \; du \text{ and } x = \frac{u^2 - b}{a}.$$

(c) Integration by Parts

$$\int u \frac{dv}{dx} \; dx = uv - \int v \frac{du}{dx} \; dx.$$

$$\left(\text{Or, more briefly, } \int u \; dv = uv - \int v \; du, \text{ where } u \text{ and } v \text{ are functions of } x. \right)$$

Choose the u and $\dfrac{dv}{dx}$ so that $v \dfrac{du}{dx}$ is easier to integrate than $u \dfrac{dv}{dx}$.

(d) Alternative Notations

$$\int f(x) \; g(x) \; dx = f(x) \int g(x) \; dx - \int \left(\int g(x) \; dx \right) f'(x) \; dx$$

$$\text{or} \quad \int uv \; dx = u \int v \; dx - \int \left(\int v \; dx \right) \frac{du}{dx} \; dx.$$

(e) Special Cases

(i) $\int f(x) \ln x \, dx$: put $u = \ln x$ and $\dfrac{dv}{dx} = f(x)$.

(ii) $\int f(x) \sin^{-1} x \, dx$: put $u = \sin^{-1} x$ and $\dfrac{dv}{dx} = f(x)$.

(iii) $\int f(x) \tan^{-1} x \, dx$: put $u = \tan^{-1} x$ and $\dfrac{dv}{dx} = f(x)$.

(iv) $\int e^{ax} \cos bx \, dx$ or $\int e^{ax} \sin bx \, dx$; integration by parts will have to be performed twice.

17.2 Worked Examples

17.1

(a) Evaluate $\displaystyle\int_0^{\pi/2} x \sin x \, dx$ correct to three significant figures.

(b) Using the substitution $u = e^x$, or otherwise,

evaluate $\displaystyle\int_0^1 \dfrac{dx}{3 - e^x}$ correct to three significant figures.

● (a) By parts.

Let $u = x$, $\dfrac{dv}{dx} = \sin x$, then $\dfrac{du}{dx} = 1$, $v = -\cos x$.

$$\int_0^{\pi/2} x \sin x \, dx = \left[-x \cos x \right]_0^{\pi/2} + \int_0^{\pi/2} \cos x \, dx$$

$$= \left[\sin x \right]_0^{\pi/2} = 1.$$

(b) If $u = e^x$ then $\dfrac{du}{dx} = e^x = u$.

When $x = 0$, $u = 1$, when $x = 1$, $u = e$.

$$\int_0^1 \frac{1}{3 - e^x} \, dx = \int_1^e \frac{1}{3 - e^x} \frac{dx}{du} \, du = \int_1^e \frac{1}{3 - u} \frac{1}{u} \, du.$$

Now $$\frac{1}{u(3-u)} = \frac{1}{3}\left\{ \frac{1}{u} + \frac{1}{3-u} \right\}$$

so $$\int_1^e \frac{1}{u(3-u)} \, du = \frac{1}{3} \int_1^e \left\{ \frac{1}{u} + \frac{1}{3-u} \right\} du$$

$$= \tfrac{1}{3} \left[\ln |u| - \ln |3 - u| \right]_1^e$$

$$= \tfrac{1}{3} (\ln |e| - \ln |3 - e|) - \tfrac{1}{3} (\ln |1| - \ln |2|)$$

$$= \frac{1}{3}\left\{ 1 + \ln\left(\frac{2}{3 - e}\right)\right\} = 0.987.$$

17.2

(a) Find $\displaystyle\int \frac{\ln x}{\sqrt{x}}\, dx$.

(b) Using the substitution $2x = \sin u$, or otherwise, evaluate $\displaystyle\int_0^{1/2} \sqrt{(1 - 4x^2)}\, dx$.

● (a) $\displaystyle\int \frac{\ln x}{\sqrt{x}}\, dx = \int x^{-1/2}\, \ln x\, dx$.

By parts, $u = \ln x$, $\dfrac{dv}{dx} = x^{-1/2}$; $\dfrac{du}{dx} = \dfrac{1}{x}$, $v = 2x^{1/2}$.

$$\int \frac{\ln x}{\sqrt{x}}\, dx = 2x^{1/2}\, \ln x - \int \left(\frac{1}{x}\right)(2x^{1/2})\, dx$$

$$= 2x^{1/2}\, \ln x - 2\int x^{-1/2}\, dx = 2x^{1/2}\, \ln x - 4x^{1/2} + C.$$

(b) If $2x = \sin u$ then $\dfrac{dx}{du} = \dfrac{1}{2} \cos u$.

When $x = 0$, $u = 0$; $x = \dfrac{1}{2}$, $u = \dfrac{\pi}{2}$.

$$\int_0^{1/2} \sqrt{(1 - 4x^2)}\, dx = \int_0^{\pi/2} \sqrt{(1 - 4x^2)}\, \frac{dx}{du}\, du$$

$$= \int_0^{\pi/2} \sqrt{(\cos^2 u)}\left(\frac{1}{2} \cos u\right) du$$

$$= \frac{1}{2}\int_0^{\pi/2} \cos^2 u\, du = \frac{1}{4}\int_0^{\pi/2} (1 + \cos 2u)\, du$$

$$= \frac{1}{4}\left[u + \frac{\sin 2u}{2}\right]_0^{\pi/2} = \left(\frac{1}{4}\right)\left(\frac{\pi}{2}\right) = \left(\frac{\pi}{8}\right).$$

17.3

(a) Find $\displaystyle\int \frac{x^3 + 1}{x^2 + 4}\, dx$.

(b) Using the substitution $t = \tan \theta$, or otherwise, evaluate $\displaystyle\int_0^{\pi/4} \frac{1}{5 + 4 \cos 2\theta}\, d\theta$,

giving your answer to three significant figures.

● (a) $\dfrac{x^3 + 1}{x^2 + 4} = x + \dfrac{(-4x + 1)}{x^2 + 4} = x - \dfrac{4x}{x^2 + 4} + \dfrac{1}{x^2 + 4}$.

$$\int \frac{x^3 + 1}{x^2 + 4}\, dx = \int \left(x - \frac{4x}{x^2 + 4} + \frac{1}{x^2 + 4} \right) dx$$

$$= \left\{ \frac{x^2}{2} - 2 \ln (x^2 + 4) + \frac{1}{2} \tan^{-1} \left(\frac{x}{2} \right) \right\} + C.$$

(b) If $t = \tan \theta$ then $\cos 2\theta = \dfrac{1 - t^2}{1 + t^2}$, $\dfrac{dt}{d\theta} = \sec^2 \theta = 1 + t^2$.

When $\theta = 0$, $t = 0$; $\theta = \dfrac{\pi}{4}$, $t = 1$.

$$\int_0^{\pi/4} \frac{1}{5 + 4 \cos 2\theta}\, d\theta = \int_0^1 \frac{1}{5 + 4 \cos 2\theta} \frac{d\theta}{dt}\, dt$$

$$= \int_0^1 \frac{1}{5 + \dfrac{4(1 - t^2)}{1 + t^2}} \frac{1}{(1 + t^2)}\, dt$$

$$= \int_0^1 \frac{1}{9 + t^2}\, dt$$

$$= \left[\frac{1}{3} \tan^{-1} \left(\frac{t}{3} \right) \right]_0^1 = \frac{1}{3} \tan^{-1} \left(\frac{1}{3} \right) = 0.107.$$

17.4

(a) Using the substitution $u^2 = x^2 + 4x$, or otherwise, evaluate $\displaystyle\int_2^4 \frac{x + 2}{\sqrt{(x^2 + 4x)}}\, dx$,

giving your answer to two significant figures.

(b) Find the exact value of $\displaystyle\int_0^2 x\sqrt{(9 - 2x^2)}\, dx$.

• (a) If $u^2 = x^2 + 4x$ then $2u \dfrac{du}{dx} = 2x + 4$, so $u = (x + 2) \dfrac{dx}{du}$.

When $x = 2$, $u^2 = 12 \implies u = 2\sqrt{3}$;

When $x = 4$, $u^2 = 32 \implies u = 4\sqrt{2}$.

$$\int_2^4 \frac{x + 2}{\sqrt{(x^2 + 4x)}}\, dx = \int_{2\sqrt{3}}^{4\sqrt{2}} \frac{1}{\sqrt{(x^2 + 4x)}} (x + 2) \left(\frac{dx}{du} \right) du$$

$$= \int_{2\sqrt{3}}^{4\sqrt{2}} \left(\frac{1}{u} \right) (u)\, du = \left[u \right]_{2\sqrt{3}}^{4\sqrt{2}}$$

$$= 4\sqrt{2} - 2\sqrt{3} = 2.2.$$

(b) Let $9 - 2x^2 = u^2$; then $-4x \dfrac{dx}{du} = 2u \implies \left(x \dfrac{dx}{du} \right) = -\dfrac{u}{2}$.

When $x = 0$, $u = 3$; $x = 2$, $u = 1$.

$$\int_0^2 x\sqrt{(9 - 2x^2)}\, dx = \int_3^1 \sqrt{(9 - 2x^2)} \left(x \frac{dx}{du} \right) du$$

$$= \int_3^1 u \left(-\frac{u}{2} \right) du = \left[-\frac{u^3}{6} \right]_3^1$$

$$= -\frac{1}{6} + \frac{27}{6} = \frac{13}{3}.$$

Notice the technique of substituting u^2 when a square root is involved.

17.5

(a) Evaluate $\int_0^1 x^2 e^{-x} dx$.

(b) Using the substitution, $x = a \tan \theta$, prove that

$$\int \frac{dx}{x^2 + a^2} = \frac{1}{a} \tan^{-1} \left(\frac{x}{a} \right).$$

● Hence, or otherwise, evaluate $\int_{-2}^0 \frac{dx}{x^2 + 4x + 8}$.

(a) By parts. Let $u = x^2$, $\frac{dv}{dx} = e^{-x}$; $\frac{du}{dx} = 2x$, $v = -e^{-x}$.

$$\int_0^1 x^2 e^{-x} dx = \left[x^2 (-e^{-x}) \right]_0^1 + \int_0^1 (e^{-x})(2x) dx$$

$$= -e^{-1} + 2 \int_0^1 x e^{-x} dx.$$

Now let $u = x$, $\frac{dv}{dx} = e^{-x}$; $\frac{du}{dx} = 1$, $v = -e^{-x}$.

$$\int_0^1 x e^{-x} dx = \left[-x e^{-x} \right]_0^1 + \int_0^1 1 e^{-x} dx$$

$$= -e^{-1} - \left[e^{-x} \right]_0^1 = -e^{-1} - e^{-1} + e^0.$$

Hence $\int_0^1 x^2 e^{-x} dx = -e^{-1} + 2(-2e^{-1} + e^0)$

$$= 2 - 5e^{-1} = 0.161.$$

(b) If $x = a \tan \theta$, $\frac{dx}{d\theta} = a \sec^2 \theta$ and $x^2 + a^2 = a^2 (\tan^2 \theta + 1) = a^2 \sec^2 \theta$.

$$\int \frac{dx}{x^2 + a^2} = \int \left(\frac{1}{x^2 + a^2} \right) \left(\frac{dx}{d\theta} \right) d\theta$$

$$= \int \frac{1}{a^2 \sec^2 \theta} a \sec^2 \theta \, d\theta = \int \frac{1}{a} d\theta$$

$$= \frac{1}{a} \theta = \frac{1}{a} \tan^{-1} \left(\frac{x}{a} \right).$$

$$\frac{1}{x^2 + 4x + 8} = \frac{1}{(x + 2)^2 + 4} = \frac{1}{(x + 2)^2 + 2^2}.$$

$$\int_{-2}^{0} \frac{1}{(x+2)^2 + 2^2}\, dx = \left[\frac{1}{2}\ \tan^{-1}\left(\frac{x+2}{2}\right)\right]_{-2}^{0}$$

$$= \frac{1}{2}\ \tan^{-1}(1) - \frac{1}{2}\ \tan^{-1}(0) = \frac{\pi}{8}.$$

17.6

(a) Evaluate $\displaystyle\int_{0}^{1} x^2\, \tan^{-1} x\, dx$, giving the answer to two significant figures.

(b) Evaluate $\displaystyle\int_{1}^{2} \frac{x^3}{\sqrt{(4-x^2)}}\, dx$.

● (a) By parts. Let $u = \tan^{-1} x$, $\dfrac{dv}{dx} = x^2$; $\dfrac{du}{dx} = \dfrac{1}{1+x^2}$, $v = \dfrac{x^3}{3}$.

$$\int_{0}^{1} x^2\, \tan^{-1} x\, dx = \left[\frac{x^3}{3}\ \tan^{-1} x\right]_{0}^{1} - \int_{0}^{1} \frac{x^3}{3}\ \frac{1}{1+x^2}\, dx$$

$$= \left(\frac{1}{3}\right)\left(\frac{\pi}{4}\right) - \frac{1}{3}\int_{0}^{1}\left(x - \frac{x}{1+x^2}\right)\, dx$$

$$= \frac{\pi}{12} - \frac{1}{3}\left[\frac{x^2}{2} - \frac{1}{2}\ \ln(1+x^2)\right]_{0}^{1}$$

$$= \frac{\pi}{12} - \frac{1}{6} + \frac{1}{6}\ \ln 2 = 0.21.$$

(b) Let $u^2 = 4 - x^2$, then $2u\dfrac{du}{dx} = -2x \Rightarrow x\dfrac{dx}{du} = -u$.

When $x = 1$, $u = \sqrt{3}$; when $x = 2$, $u = 0$.

$$\int_{1}^{2} \frac{x^3}{\sqrt{(4-x^2)}}\, dx = \int_{\sqrt{3}}^{0} \frac{x^2}{\sqrt{(4-x^2)}}\left(x\frac{dx}{du}\right)du$$

$$= -\int_{\sqrt{3}}^{0} \frac{(4-u^2)}{u}\ (u\, du) = \left[\frac{u^3}{3} - 4u\right]_{\sqrt{3}}^{0}$$

$$= -\frac{3\sqrt{3}}{3} + 4\sqrt{3} = 3\sqrt{3} = 5.20.$$

17.7

(a) Evaluate $\displaystyle\int_{3}^{7} \frac{1}{\sqrt{(7+6x-x^2)}}\, dx$.

(b) Using the substitution $u^2 = 3 - x$, or otherwise, find $\displaystyle\int (x+1)\sqrt{(3-x)}\, dx$.

● (a) $7 + 6x - x^2 = -(x^2 - 6x - 7) = -\{(x-3)^2 - 16\}$.

Thus $\sqrt{(7 + 6x - x^2)} = \sqrt{\{16 - (x - 3)^2\}}$.

Let $x - 3 = 4 \sin u$, then $\dfrac{dx}{du} = 4 \cos u$ and

$$16 - (x - 3)^2 = 16 - 16 \sin^2 u = 16 \cos^2 u.$$

When $x = 3$, $u = 0$; when $x = 7$, $u = \dfrac{\pi}{2}$.

$$\int_3^7 \dfrac{1}{\sqrt{(7 + 6x - x^2)}} \, dx = \int_0^{\pi/2} \dfrac{1}{\sqrt{\{16 - (x - 3)^2\}}} \left(\dfrac{dx}{du}\right) du$$

$$= \int_0^{\pi/2} \left(\dfrac{1}{4 \cos u}\right) (4 \cos u) \, du$$

$$= \left[u\right]_0^{\pi/2} = \dfrac{\pi}{2}.$$

(b) Let $u^2 = 3 - x$, then $2u \dfrac{du}{dx} = -1$ or $-2u = \dfrac{dx}{du}$.

$$\int (x + 1)\sqrt{(3 - x)} \left(\dfrac{dx}{du}\right) du = \int (4 - u^2)(u)(-2u) \, du$$

$$= -2 \int (4u^2 - u^4) \, du$$

$$= -2 \left(\dfrac{4}{3} u^3 - \dfrac{1}{5} u^5\right) = -\dfrac{2}{15} u^3 (20 - 3u^2)$$

$$= \dfrac{-2}{15} (3 - x)^{3/2} (20 - 9 + 3x)$$

$$= \dfrac{-2}{15} (3 - x)^{3/2} (11 + 3x) + C.$$

17.8

(a) Evaluate $\displaystyle\int_0^1 x^3 e^{x^2} \, dx$.

(b) Using the substitution $t = \tan \dfrac{\theta}{2}$, or otherwise,

evaluate $\displaystyle\int_0^{\pi/2} \dfrac{6}{1 + \sin \theta + 3 \cos \theta} \, d\theta$.

● (a) By parts. Since $\dfrac{d}{dx} (e^{x^2}) = 2x e^{x^2}$, choose $\dfrac{dv}{dx} = x e^{x^2}$ and $u = x^2$.

Then $\dfrac{du}{dx} = 2x$, $v = \dfrac{1}{2} e^{x^2}$.

$$\int_0^1 x^3 e^{x^2} \, dx = \left[\dfrac{1}{2} x^2 e^{x^2}\right]_0^1 - \dfrac{1}{2} \int_0^1 2x e^{x^2} \, dx$$

$$= \left[\dfrac{1}{2} x^2 e^{x^2} - \dfrac{1}{2} e^{x^2}\right]_0^1$$

$$= (\tfrac{1}{2} e - \tfrac{1}{2} e) - (0 - \tfrac{1}{2}) = \tfrac{1}{2}.$$

(b) Let $t = \tan \dfrac{\theta}{2}$, then $\dfrac{dt}{d\theta} = \tfrac{1}{2}\sec^2 \dfrac{\theta}{2} = \tfrac{1}{2}(1 + t^2)$ $\Rightarrow$ $\dfrac{d\theta}{dt} = \dfrac{2}{(1 + t^2)}$.

When $\theta = 0$, $t = 0$; when $\theta = \dfrac{\pi}{2}$, $t = 1$.

$$\int_0^1 \frac{6}{1 + \sin\theta + 3\cos\theta} \; \frac{d\theta}{dt} \; dt$$

$$= \int_0^1 \left(\frac{6}{1 + \dfrac{2t}{1 + t^2} + 3\dfrac{(1 - t^2)}{1 + t^2}} \right) \frac{2}{1 + t^2} \; dt$$

$$= \int_0^1 \frac{12}{1 + t^2 + 2t + 3 - 3t^2} \; dt$$

$$= \int_0^1 \frac{6}{2 + t - t^2} \; dt.$$

By partial fractions, $\dfrac{6}{(2 - t)(1 + t)} = \dfrac{2}{2 - t} + \dfrac{2}{1 + t}$.

$$\int_0^1 \frac{6}{2 + t - t^2} \; dt = 2 \int_0^1 \left(\frac{1}{2 - t} + \frac{1}{1 + t} \right) dt$$

$$= 2 \left[-\ln|2 - t| + \ln|1 + t| \right]_0^1$$

$$= 2(-\ln 1 + \ln 2) - 2(-\ln 2 + \ln 1)$$

$$= 4 \ln 2$$

$$= 2.77.$$

17.9

(a) Evaluate $\displaystyle\int_1^3 x^3 \ln x \, dx$.

(b) Calculate $\displaystyle\int_0^4 \frac{5x^2 + 4x + 17}{(2x + 1)(16 + x^2)} \, dx$,

leaving your answer in exact form.

● (a) By parts, $u = \ln x$, $\dfrac{dv}{dx} = x^3$, $\dfrac{du}{dx} = \dfrac{1}{x}$, $v = \dfrac{x^4}{4}$.

$$\int_1^3 x^3 \ln x \, dx = \left[\left(\frac{x^4}{4}\right)(\ln x) \right]_1^3 - \int_1^3 \left(\frac{x^4}{4}\right)\left(\frac{1}{x}\right) dx$$

$$= \left[\frac{x^4}{4} \ln x - \frac{x^4}{16} \right]_1^3$$

$$= \tfrac{81}{4}\ln 3 - \tfrac{81}{16} - \tfrac{1}{4}\ln 1 + \tfrac{1}{16}$$

$$= \tfrac{81}{4}\ln 3 - 5 = 17.2.$$

(b) $\dfrac{5x^2 + 4x + 17}{(2x + 1)(16 + x^2)} \equiv \dfrac{A}{2x + 1} + \dfrac{Bx + C}{16 + x^2}$

$$\equiv \dfrac{A(16 + x^2) + (Bx + C)(2x + 1)}{(2x + 1)(16 + x^2)}.$$

Equating numerators:

Let $x = -\frac{1}{2}$, $\frac{5}{4} - 2 + 17 = A(16 + \frac{1}{4})$ $\Rightarrow$ $A = 1$.

Let $x = 0$, $17 = A(16) + C(1)$ $\Rightarrow$ $C = 1$.

Let $x = 1$, $5 + 4 + 17 = A(17) + (B + C)3$ $\Rightarrow$ $B = 2$.

$$\int_0^4 \dfrac{5x^2 + 4x + 17}{(2x + 1)(16 + x^2)}\, dx = \int_0^4 \left(\dfrac{1}{2x + 1} + \dfrac{2x + 1}{16 + x^2} \right) dx$$

$$= \int_0^4 \left(\dfrac{1}{2x + 1} + \dfrac{2x}{16 + x^2} + \dfrac{1}{16 + x^2} \right) dx$$

$$= \left[\dfrac{1}{2} \ln |2x + 1| + \ln |16 + x^2| + \dfrac{1}{4} \tan^{-1} \dfrac{x}{4} \right]_0^4$$

$$= (\tfrac{1}{2} \ln 9 + \ln 32 + \tfrac{1}{4} \tan^{-1} 1) - (\ln 16)$$

$$= \ln 3 + \ln 2 + \dfrac{\pi}{16} = \ln 6 + \dfrac{\pi}{16}$$

$$= 1.99\,.$$

17.3 Exercises

17.1 Find   (a) $\displaystyle\int \dfrac{x^2}{\sqrt{(1 - x^2)}}\, dx$;   (b) $\displaystyle\int \dfrac{x}{\sqrt{(1 - x^2)}}\, dx$.

17.2 Find   (a) $\displaystyle\int \dfrac{x^2}{1 + x^2}\, dx$;   (b) $\displaystyle\int \dfrac{x^2}{1 - x^2}\, dx$.

17.3

(a) Evaluate $\displaystyle\int_{\pi/2}^{\pi} x \cos x\, dx$, correct to three significant figures.

(b) By using the substitution $u = e^x$, or otherwise, evaluate

$$\int_0^1 \dfrac{e^{3x}}{1 + e^{2x}}\, dx.$$

17.4

(a) Find $\displaystyle\int \dfrac{\ln x}{x^2}\, dx$.

(b) By using the substitution $t = \tan \dfrac{x}{2}$, evaluate

$$\int_0^{\pi/2} \dfrac{1}{4 + 5 \cos x}\, dx,\quad \text{giving your answer to three significant figures.}$$

17.5

(a) Using the substitution $x = \tan u$, or otherwise, evaluate

$$\int_0^1 \frac{dx}{(1 + x^2)^2} .$$

(b) Find $\displaystyle\int_0^{1/2} \frac{x + 2}{4x^2 + 1}\, dx.$

17.6

(a) Evaluate $\displaystyle\int_0^1 (x + 1)\, e^{2x}\, dx.$

(b) Find $\displaystyle\int x^2 \sin^{-1} x\, dx.$

17.7

(a) Evaluate $\displaystyle\int_4^7 \frac{1}{x^2 - 8x + 25}\, dx.$

(b) Find $\displaystyle\int \frac{x + 3}{\sqrt{(2x - 7)}}\, dx.$

17.8

(a) By means of the substitution $t = \tan x$, or otherwise, evaluate

$$\int_0^{\pi/4} \frac{dx}{1 + \cos^2 x} .$$

(b) Find $\displaystyle\int (2 + x)\, (2 - x)^{3/2}\, dx.$

17.9

(a) Find $\displaystyle\int \sin\theta\, \sqrt{(1 + \cos\theta)}\, d\theta.$

(b) Evaluate $\displaystyle\int_0^{\pi/3} \frac{\sin\theta}{1 + \cos 2\theta}\, d\theta.$

17.4 Outline Solutions to Exercises

17.1

(a) $x = \sin u \quad\Rightarrow\quad \dfrac{dx}{du} = \cos u.$

$$\int \frac{x^2}{\sqrt{(1 - x^2)}}\, dx = \int \sin^2 u\, du = \frac{1}{2}\left[u - \frac{\sin 2u}{2} \right]$$

$$= \tfrac{1}{2}\{\sin^{-1} x - x\sqrt{(1 - x^2)}\} + C.$$

(b) $u^2 = 1 - x^2 \Rightarrow 2u \dfrac{du}{dx} = -2x \Rightarrow x \dfrac{dx}{du} = -u.$

$$\int \frac{x}{\sqrt{(1 - x^2)}}\, dx = \int -1 \, du = -u = -\sqrt{(1 - x^2)} + C.$$

17.2

(a) $x = \tan u \Rightarrow \dfrac{dx}{du} = \sec^2 u.$

$$\int \frac{x^2}{1 + x^2}\, dx = \int \tan^2 u \, du = \int (\sec^2 u - 1)\, du$$

$$= \tan u - u = x - \tan^{-1} x + C.$$

(b) $\displaystyle \int \frac{x^2}{1 - x^2}\, dx = \int \left(-1 + \frac{1}{1 - x^2}\right) dx = \frac{1}{2}\int \left(-2 + \frac{1}{1 - x} + \frac{1}{1 + x}\right) dx$

$$= -x - \tfrac{1}{2} \ln |1 - x| + \tfrac{1}{2} \ln |1 + x| + C$$

$$= -x + \tfrac{1}{2} \ln \left|\frac{1 + x}{1 - x}\right| + C.$$

17.3

(a) By parts, $u = x$, $\dfrac{dv}{dx} = \cos x$ gives $\Big[x \sin x + \cos x \Big]_{\pi/2}^{\pi}$

$$\Rightarrow \quad -\left(1 + \frac{\pi}{2}\right).$$

(b) $u = e^x \Rightarrow \dfrac{du}{dx} = e^x = u.$

$$\int_0^1 \frac{e^{3x}}{1 + e^{2x}}\, dx = \int_1^e \frac{u^2}{1 + u^2}\, du$$

$$= \int_1^e \left(1 - \frac{1}{1 + u^2}\right) du = \Big[u - \tan^{-1} u \Big]_1^e = 1.29.$$

17.4

(a) By parts, $u = \ln x$, $\dfrac{dv}{dx} = \dfrac{1}{x^2}$ leads to $-\dfrac{1}{x} \ln x + \displaystyle\int \frac{1}{x^2}\, dx$

$$= -\frac{1}{x}(1 + \ln x) + C.$$

(b) $t = \tan \dfrac{x}{2}$ so $\dfrac{dt}{dx} = \dfrac{1}{2} \sec^2 \dfrac{x}{2} = \dfrac{1 + t^2}{2} \Rightarrow \dfrac{dx}{dt} = \dfrac{2}{1 + t^2}.$

$$\int_0^{\pi/2} \frac{1}{4 + 5 \cos x}\, dx = \int_0^1 \frac{2}{9 - t^2}\, dt = \frac{1}{3}\int_0^1 \left(\frac{1}{3 - t} + \frac{1}{3 + t}\right) dt$$

$$= \tfrac{1}{3}\Big[-\ln |3 - t| + \ln |3 + t| \Big]_0^1 = \tfrac{1}{3} \ln 2 = 0.231.$$

17.5

(a) $x = \tan u \Rightarrow \dfrac{dx}{du} = \sec^2 u = 1 + x^2$.

$$\int_0^1 \frac{dx}{(1 + x^2)^2} = \int_0^{\pi/4} \cos^2 u\, du = \tfrac{1}{2} \int_0^{\pi/4} (1 + \cos 2u)\, du = \tfrac{1}{8}(\pi + 2).$$

(b) $$\int_0^{1/2} \frac{x}{4x^2 + 1}\, dx = \tfrac{1}{8}\Big[\ln(4x^2 + 1)\Big]_0^{1/2} = \tfrac{1}{8}\ln 2. \qquad (1)$$

For $\displaystyle\int_0^{1/2} \frac{2}{4x^2 + 1}\, dx$, let $2x = \tan u$, $\;2\dfrac{dx}{du} = \sec^2 u$,

$$\Rightarrow \int_0^{\pi/4} (1)\, du = \Big[u\Big]_0^{\pi/4} = \frac{\pi}{4}. \qquad (2)$$

Adding equations 1 and 2 gives $\dfrac{1}{8}\ln 2 + \dfrac{\pi}{4} = 0.872$.

17.6

(a) By parts, $u = x + 1$, $\dfrac{dv}{dx} = e^{2x}$, gives

$$\Big[\tfrac{1}{2}(x + 1)e^{2x} - \tfrac{1}{4}e^{2x}\Big]_0^1 = \tfrac{3}{4}e^2 - \tfrac{1}{4}e^0 = 5.29.$$

(b) By parts, $u = \sin^{-1} x$, $\dfrac{dv}{dx} = x^2$, gives

$$\frac{x^3}{3}\sin^{-1} x - \int\left(\frac{x^3}{3}\right)\frac{1}{\sqrt{(1 - x^2)}}\, dx.$$

Let $1 - x^2 = w^2$, then $-2x\dfrac{dx}{dw} = 2w \Rightarrow x\dfrac{dx}{dw} = -w$.

$$\int\frac{x^3\, dx}{3\sqrt{(1 - x^2)}} = \frac{1}{3}\int\frac{(1 - w^2)}{w}(-w)\, dw = -\frac{1}{3}\int(1 - w^2)\, dw$$

$$= -\frac{w}{9}(3 - w^2) = -\frac{1}{9}(2 + x^2)\sqrt{(1 - x^2)}.$$

Hence $\displaystyle\int x^2 \sin^{-1} x\, dx = \frac{x^3}{3}\sin^{-1} x + \left(\frac{2 + x^2}{9}\right)\sqrt{(1 - x^2)} + C$.

17.7

(a) $x^2 - 8x + 25 = (x - 4)^2 + 9 \Rightarrow$ put $x - 4 = 3\tan u$, $\dfrac{dx}{du} = 3\sec^2 u$.

$$\int_4^7 \frac{1}{x^2 - 8x + 25}\, dx = \int_0^{\pi/4} \tfrac{1}{3}\, du = \Big[\tfrac{1}{3}u\Big]_0^{\pi/4} = \frac{\pi}{12}.$$

(b) Let $2x - 7 = u^2$, $\;2\dfrac{dx}{du} = 2u$.

$$\int \frac{x+3}{\sqrt{(2x-7)}} \, dx = \int \frac{(u^2+13)}{2} \, du = \frac{1}{2} \left(\frac{u^3}{3} + 13u \right) = \frac{u}{6} (u^2+39)$$

$$= \tfrac{1}{3}(x+16)\sqrt{(2x-7)} + C.$$

17.8

(a) $t = \tan x$, $\dfrac{dt}{dx} = 1 + t^2$ so $\dfrac{dx}{dt} = \dfrac{1}{1+t^2}$.

$$\int_0^{\pi/4} \frac{dx}{1+\cos^2 x} = \int_0^1 \frac{1}{2+t^2} \, dt = \left[\frac{1}{\sqrt{2}} \tan^{-1} \frac{t}{\sqrt{2}} \right]_0^1 = 0.435.$$

(b) Let $2 - x = u^2$, $-\dfrac{dx}{du} = 2u$.

$$\int (2+x)(2-x)^{3/2} \, dx = -2 \int (4u^4 - u^6) \, du = -\frac{2u^5}{35} (28 - 5u^2)$$

$$= -\tfrac{2}{35} (18 + 5x)(2-x)^{5/2}.$$

17.9

(a) Let $1 + \cos\theta = u^2$, $-\sin\theta \dfrac{d\theta}{du} = 2u$.

$$\int \sin\theta \sqrt{(1+\cos\theta)} \, d\theta = \int -2u^2 \, du = -\tfrac{2}{3}(1+\cos\theta)^{3/2} + C.$$

(b) $\displaystyle\int_0^{\pi/3} \frac{\sin\theta}{2\cos^2\theta} \, d\theta = \tfrac{1}{2} \int_0^{\pi/3} \sec\theta \tan\theta \, d\theta = \left[\tfrac{1}{2}\sec\theta \right]_0^{\pi/3} = 0.5.$

18 Applications of Integration

The idea of area under a curve as the limit of a sum of area of rectangles.
Simple applications of integration to plane areas and volumes of revolution.

18.1 Fact Sheet

(a) Area as the Limit of a Sum

The area A enclosed by the curve $y = f(x)$, $(y > 0)$, the x axis and the lines $x = a$ and $x = b$, $(a \leqslant x \leqslant b)$, may be divided into n strips, each of width δx, height $y = f(x)$, area $\delta A \approx y\delta x \approx f(x)\,\delta x$.

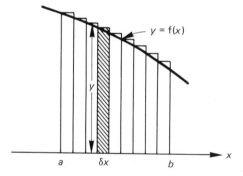

Summing for all the strips,

$$\sum \delta A \approx \sum_{n} f(x)\,\delta x.$$

If n is increased and δx decreased, in the limit as n tends to infinity,

$$\text{area } A = \int_{a}^{b} f(x)\,\mathrm{d}x.$$

(b) Calculation of Areas

- ### *Area between a curve and the x-axis*

(i) If $y = f(x) \geqslant 0$ for $a \leqslant x \leqslant b$, area of elemental strip $= y \, \delta x \approx f(x) \, \delta x$.

$$\text{Area} = \int_a^b y \, dx = \int_a^b f(x) \, dx.$$

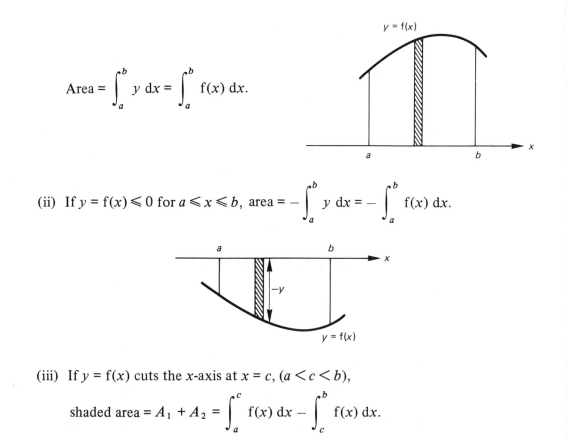

(ii) If $y = f(x) \leqslant 0$ for $a \leqslant x \leqslant b$, area $= -\int_a^b y \, dx = -\int_a^b f(x) \, dx.$

(iii) If $y = f(x)$ cuts the x-axis at $x = c$, $(a < c < b)$,

$$\text{shaded area} = A_1 + A_2 = \int_a^c f(x) \, dx - \int_c^b f(x) \, dx.$$

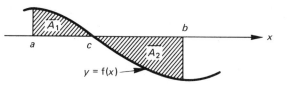

- ### *Area between two curves*

If $y_1 = f_1(x)$ and $y_2 = f_2(x)$ intersect when $x = a$ and $x = b$, and if $f_1(x) > f_2(x)$ for $a < x < b$, area of elemental strip $\approx (y_1 - y_2) \, \delta x = [f_1(x) - f_2(x)] \, \delta x$.

$$\text{Area} = \int_a^b [f_1(x) - f_2(x)] \, dx.$$

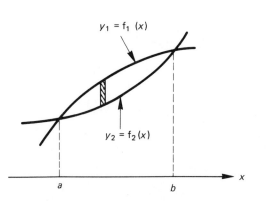

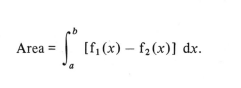

● *Area between a curve and the y-axis*

If $x = g(y) \geqslant 0$ for $c \leqslant y \leqslant d$, area of elemental strip $\approx x\,\delta y = g(y)\,\delta y$.

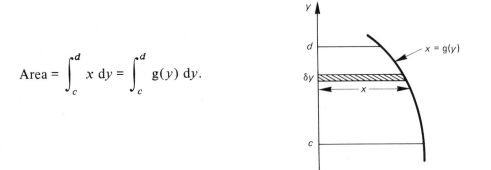

$$\text{Area} = \int_c^d x\,\mathrm{d}y = \int_c^d g(y)\,\mathrm{d}y.$$

(c) Calculation of Volumes of Revolution

(i) *About the x-axis.* If $y = f(x)$, for $a \leqslant x \leqslant b$, volume of elemental disc $\approx \pi\,y^2\,\delta x$.

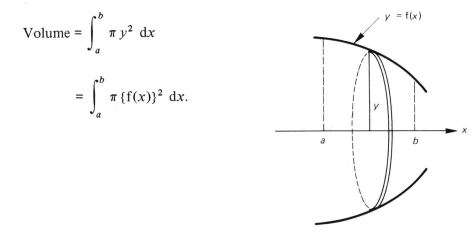

$$\text{Volume} = \int_a^b \pi\,y^2\,\mathrm{d}x$$

$$= \int_a^b \pi\,\{f(x)\}^2\,\mathrm{d}x.$$

(ii) *About the y-axis.* If $x = g(y)$ for $c \leqslant y \leqslant d$, volume of elemental disc $\approx \pi\,x^2\,\delta y$.

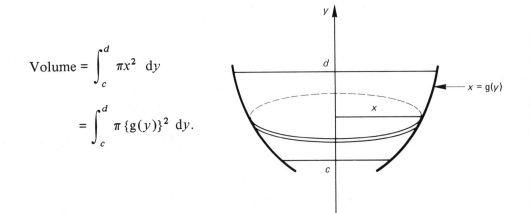

$$\text{Volume} = \int_c^d \pi x^2\,\mathrm{d}y$$

$$= \int_c^d \pi\,\{g(y)\}^2\,\mathrm{d}y.$$

(iii) *About the line* y = c (*parallel to the* x-*axis*).
    If $y = f(x)$, volume of elemental disc $\approx \pi\,(y - c)^2\ \delta x$.

Volume $= \displaystyle\int_{a}^{b} \pi\,(y - c)^2\ \mathrm{d}x$

$\quad\quad\ = \displaystyle\int_{a}^{b} \pi\,\{f(x) - c\}^2\ \mathrm{d}x.$

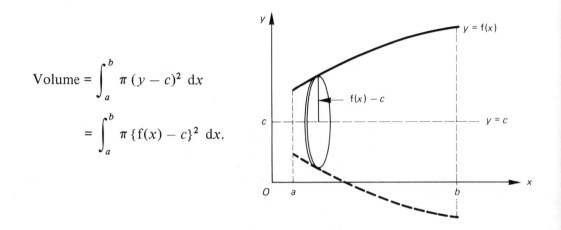

(iv) *About the line* x = a (*parallel to the* y-*axis*).
    If $x = g(y)$, volume of elemental disc $\approx \pi\,(x - a)^2\ \delta y$.

Volume $= \displaystyle\int_{c}^{d} \pi\,(x - a)^2\ \mathrm{d}y$

$\quad\quad\ = \displaystyle\int_{c}^{d} \pi\,\{g(y) - a\}^2\ \mathrm{d}y.$

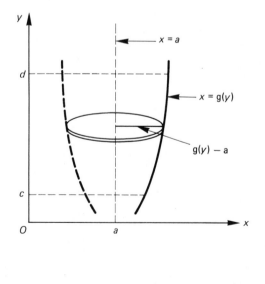

18.2 Worked Examples

18.1 Calculate the area of the finite region bounded by the curve $y = x (6 - x)$ and the straight line $y = 2x$.

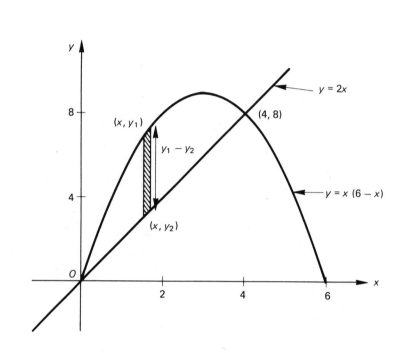

The two curves intersect when $2x = x (6 - x)$ $\Rightarrow$ $x = 0$ or 4.
Area of shaded element $\approx (y_1 - y_2)\, \delta x$,
where $y_1 = x (6 - x)$ and $y_2 = 2x$.

$$\text{Area of the finite region} = \int_0^4 \{(6x - x^2) - 2x\}\, dx$$

$$= \int_0^4 (4x - x^2)\, dx$$

$$= \left[2x^2 - \frac{x^3}{3}\right]_0^4$$

$$= 32 - \tfrac{64}{3}.$$

Area of the finite region = $\tfrac{32}{3}$ square units.

18.2 Find the area enclosed by the curve $y^2 = 4ax$ and the straight line $x = 3a$.

Find the volume of the solid of revolution formed when the region is rotated through four right angles about the line $x = 3a$.

- Curve $y^2 = 4ax$ and line $x = 3a$ intersect when

$$y^2 = (4a)(3a) \quad \Rightarrow \quad y = \pm 2a\sqrt{3}.$$

Area of shaded element $\approx (x_2 - x_1)\,\delta y$ where

$$x_1 = \frac{y^2}{4a} \quad \text{and} \quad x_2 = 3a.$$

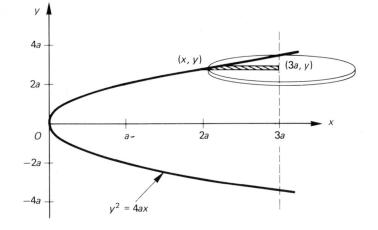

$$\text{Enclosed area} = \int_{-2a\sqrt{3}}^{2a\sqrt{3}} \left(3a - \frac{y^2}{4a}\right)dy = 2\int_{0}^{2a\sqrt{3}} \left(3a - \frac{y^2}{4a}\right)dy$$

$$= 2\left[3ay - \frac{y^3}{12a}\right]_{0}^{2a\sqrt{3}}$$

$$= 2\left[6a^2\sqrt{3} - \frac{24a^3}{12a}\sqrt{3}\right]$$

$$= 8a^2\sqrt{3} \text{ square units.}$$

Volume of disc obtained by rotating the element about the line $x = 3a \approx \pi(x_2 - x_1)^2\,\delta y$.

$$\text{Volume of solid} = 2\pi\int_{0}^{2a\sqrt{3}} \left(3a - \frac{y^2}{4a}\right)^2 dy$$

$$= 2\pi\int_{0}^{2a\sqrt{3}} \left(9a^2 - \frac{3y^2}{2} + \frac{y^4}{16a^2}\right)dy$$

$$= 2\pi\left[9a^2y - \frac{y^3}{2} + \frac{y^5}{80a^2}\right]_{0}^{2a\sqrt{3}}$$

$$= 2\pi\left(18a^3\sqrt{3} - 12a^3\sqrt{3} + \frac{18a^3\sqrt{3}}{5}\right).$$

$$\text{Volume of solid} = \frac{96\pi a^3\sqrt{3}}{5} \text{ cubic units.}$$

18.3 Draw the graphs of sin $3x$ and sin x for $0 \leqslant x \leqslant \pi$ on the same axes and calculate the points of intersection.

Find the areas of the three regions enclosed by these curves.

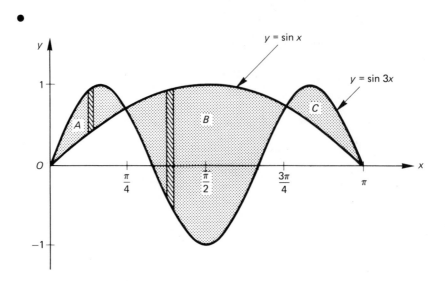

At points of intersection sin $3x = \sin x$,

$$3x = n\pi + (-1)^n x, \text{ so } x = 0, \frac{\pi}{4}, \frac{3\pi}{4}, \pi.$$

For area A: Area of shaded element $\approx (y_1 - y_2)\,\delta x$,

where $y_1 = \sin 3x$ and $y_2 = \sin x$.

$$\text{Area } A = \int_0^{\pi/4} (\sin 3x - \sin x)\,dx$$

$$= \left[\frac{-\cos 3x}{3} + \cos x \right]_0^{\pi/4}$$

$$= \left(\frac{\sqrt{2}}{6} + \frac{\sqrt{2}}{2} \right) - \left(-\frac{1}{3} + 1 \right)$$

$$= \tfrac{2}{3}\,(\sqrt{2} - 1) \text{ square units.}$$

Area C = Area A.

For area B: area of shaded element $\approx (y_2 - y_1)\,\delta x$.

$$\text{Area } B = \left[-\cos x + \frac{\cos 3x}{3} \right]_{\pi/4}^{3\pi/4}$$

$$= \left(\frac{\sqrt{2}}{2} + \frac{\sqrt{2}}{6} \right) - \left(-\frac{\sqrt{2}}{2} - \frac{\sqrt{2}}{6} \right)$$

$$= \tfrac{4}{3}\sqrt{2} \text{ square units.}$$

The enclosed areas are $\tfrac{2}{3}\,(\sqrt{2} - 1)$, $\tfrac{4}{3}\sqrt{2}$ and $\tfrac{2}{3}\,(\sqrt{2} - 1)$.

18.4 Find the volume of the solid of revolution formed when the region enclosed by the curve $y = xe^{-x}$, the x-axis and the line $x = 3$ is rotated completely about the x-axis.

● Volume of shaded disc $\approx \pi y^2\ \delta x$.

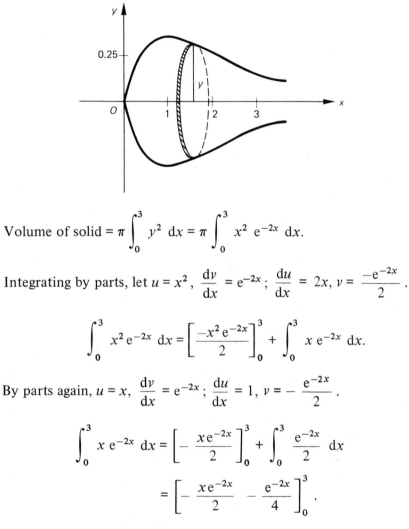

$$\text{Volume of solid} = \pi \int_0^3 y^2\ dx = \pi \int_0^3 x^2\ e^{-2x}\ dx.$$

Integrating by parts, let $u = x^2$, $\dfrac{dv}{dx} = e^{-2x}$; $\dfrac{du}{dx} = 2x,\ v = \dfrac{-e^{-2x}}{2}$.

$$\int_0^3 x^2 e^{-2x}\ dx = \left[\frac{-x^2 e^{-2x}}{2}\right]_0^3 + \int_0^3 x\,e^{-2x}\ dx.$$

By parts again, $u = x$, $\dfrac{dv}{dx} = e^{-2x}$; $\dfrac{du}{dx} = 1,\ v = -\dfrac{e^{-2x}}{2}$.

$$\int_0^3 x\,e^{-2x}\ dx = \left[-\frac{x\,e^{-2x}}{2}\right]_0^3 + \int_0^3 \frac{e^{-2x}}{2}\ dx$$

$$= \left[-\frac{x\,e^{-2x}}{2} - \frac{e^{-2x}}{4}\right]_0^3.$$

Hence the volume of revolution

$$V = \pi \left[\frac{-x^2}{2}\,e^{-2x} - \frac{x}{2}\,e^{-2x} - \frac{e^{-2x}}{4}\right]_0^3$$

$$= \pi \left\{\frac{-e^{-6}}{4}\ (18 + 6 + 1) + \frac{e^0}{4}\right\}.$$

$V = 0.737$ cubic units.

18.5 The periodic function $f(x)$ is such that

$$f(x) = \begin{cases} x, & 0 \leqslant x < 1, \\ 2 - x, & 1 \leqslant x < 2, \\ 0, & 2 \leqslant x < 3. \end{cases}$$

and $f(x + 3) = f(x)$.

Sketch the graph of f(x) for $-5 \leqslant x \leqslant 5$ and evaluate $\int_0^9 f(x)\,dx$.

● $f(x + 3) = f(x)$ indicates that the graph is periodic, period 3.

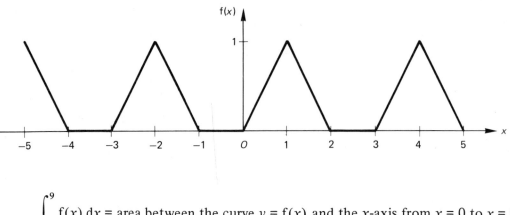

$\int_0^9 f(x)\,dx$ = area between the curve $y = f(x)$ and the x-axis from $x = 0$ to $x = 9$

= the area of three triangles.
Area of one triangle = 1 square unit.

Therefore $\int_0^9 f(x)\,dx = 3$.

18.6 Sketch the curve with equation $y^2 = x\,(x - 2)^2$. Find the area of the loop and the volume of the solid generated when this region is rotated through two right angles about the x-axis.

● Sketch.
   (a) This is an even function of y, and therefore is symmetrical about the x-axis.
   (b) $y^2 \geqslant 0$ $\Rightarrow$ $x\,(x - 2)^2 \geqslant 0$ $\Rightarrow$ $x \geqslant 0$.
   (c) Curve cuts x-axis at $x = 0$ and $x = 2$.
   (d) $y = \pm(x - 2)\sqrt{x}$.
      As $x \to 0$, y behaves as the curve $y = \mp 2\sqrt{x}$ (with infinite gradient).
      As $x \to 2$, y behaves as the curve $y = \pm (x - 2)\sqrt{2}$ (two straight lines, gradients $\pm\sqrt{2}$.)

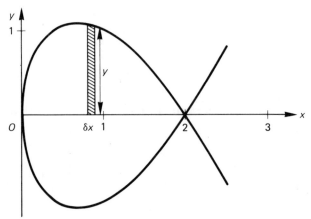

Area of shaded element $\approx y \, \delta x = x^{1/2} \, (2 - x) \, \delta x$.

$$\text{Area of loop} = 2 \int_0^2 y \, dx = 2 \int_0^2 (2x^{1/2} - x^{3/2}) \, dx$$

$$= 2 \left[\frac{4x^{3/2}}{3} - \frac{2x^{5/2}}{5} \right]_0^2$$

$$= 2 \left(\frac{8}{3} \sqrt{2} - \frac{8}{5} \sqrt{2} \right)$$

$$= \frac{32\sqrt{2}}{15} \text{ square units.}$$

Volume of disc generated by rotating the shaded element about the x-axis $\approx \pi \, y^2 \, \delta x$.

$$\text{Volume of revolution} = \pi \int_0^2 x \, (x - 2)^2 \, dx$$

$$= \pi \int_0^2 (x^3 - 4x^2 + 4x) \, dx$$

$$= \pi \left[\frac{x^4}{4} - \frac{4x^3}{3} + 2x^2 \right]_0^2$$

$$= \pi \left(4 - \frac{32}{3} + 8 \right)$$

$$= \frac{4\pi}{3} \text{ cubic units.}$$

18.7 Find the area of the region enclosed by the curve $y = 2x + 3 \cos x$, the x-axis and the lines $x = 0$ and $x = \pi$. Find the volume of the solid generated when this region is rotated completely about the x-axis.

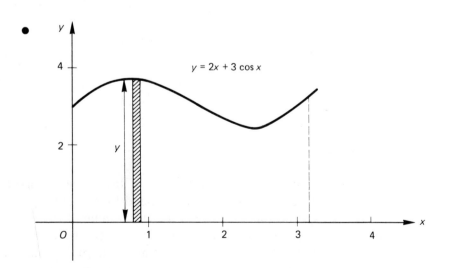

The area of the shaded element $\approx y\,\delta x$.

$$\text{Required area} = \int_0^\pi (2x + 3\cos x)\,dx$$

$$= \left[x^2 + 3\sin x\right]_0^\pi$$

$$= \pi^2$$

$$= 9.87 \text{ square units.}$$

Volume of disc generated when shaded element is rotated about the x-axis
$\approx \pi\,(2x + 3\cos x)^2\,\delta x$.

$$\text{Volume of revolution} = \pi \int_0^\pi (4x^2 + 12x\cos x + 9\cos^2 x)\,dx.$$

Now, by parts, $\displaystyle\int x\cos x\,dx = x\sin x - \int \sin x\,dx$

$$= x\sin x + \cos x.$$

Also, $\displaystyle\int(4x^2 + 9\cos^2 x)\,dx = \int\{4x^2 + \tfrac{9}{2}(1 + \cos 2x)\}\,dx$

$$= \frac{4x^3}{3} + \frac{9x}{2} + \frac{9}{4}\sin 2x.$$

Thus,

$$\text{volume of revolution} = \pi\left[\frac{4x^3}{3} + \frac{9x}{2} + \frac{9}{4}\sin 2x + 12x\sin x + 12\cos x\right]_0^\pi$$

$$= \pi\left(\frac{4\pi^3}{3} + \frac{9\pi}{2} - 12\right) - \pi(12)$$

$$= 98.9 \text{ cubic units.}$$

18.8 The inner surface of a bowl is of the shape formed by rotating completely about the y-axis the area bounded by the curve $y = x^2 - 4$, the x-axis, the y-axis and the line $y = 3$. Find the volume of the bowl.

Calculate the volume of water in the bowl when the depth of water is d (< 3). If water is poured in at a rate of 5 cubic units per second, find the rate at which the depth is increasing when $d = 1$.

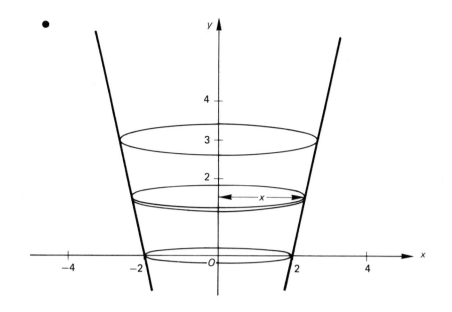

Volume of disc $\delta V \approx \pi\, x^2\; \delta y$.

Volume of bowl $V = \pi \displaystyle\int_0^3 x^2 \; \mathrm{d}y$.

But $x^2 = y + 4$, so volume of bowl $= \pi \displaystyle\int_0^3 (y + 4)\,\mathrm{d}y$

$$= \pi \left[\frac{y^2}{2} + 4y \right]_0^3$$

$$= \pi \left(\frac{9}{2} + 12 \right).$$

Volume of bowl $= \dfrac{33\pi}{2} = 51.8$ cubic units.

When the water has a depth of d,

$$\text{volume of water } V = \pi \left[\frac{y^2}{2} + 4y \right]_0^d$$

$$= \pi \left(\frac{d^2}{2} + 4d \right).$$

$\dfrac{\mathrm{d}V}{\mathrm{d}(d)} = \pi\,(d + 4)$, so, with $\dfrac{\mathrm{d}V}{\mathrm{d}t} = 5$,

$$\frac{\mathrm{d}(d)}{\mathrm{d}t} = \frac{\mathrm{d}(d)}{\mathrm{d}V} \left(\frac{\mathrm{d}V}{\mathrm{d}t} \right) = \frac{5}{\pi\,(d + 4)}.$$

Depth is increasing at a rate of $\dfrac{5}{\pi\,(d + 4)}$ units per second.

When $d = 1$, the depth is increasing at $\dfrac{1}{\pi} = 0.32$ units per second.

18.3 Exercises

18.1 Sketch the graph $y = x(x - 2)(3 - x)$. Find the equation of the tangents at $x = 2$ and $x = 3$. Find the area of the region enclosed by the curve and these two tangents.

18.2 Find the area of the region enclosed by the curves $y^2 = 4x$ and $y^2 = 8 - 4x$.
   Find the volume of the solid generated when this region is rotated about the y-axis.

18.3 By considering a circle centre $(0, -r)$, radius r, or otherwise find the volume of the cap of depth h of a sphere, radius r.

18.4 The region A is bounded by the x-axis, the curve $y = \tan x$ and the line $x = \dfrac{\pi}{3}$.
   Find
   (a) the area of A,
   (b) the volume of the solid of revolution formed when A is rotated about the x-axis.

18.5 Sketch the curve $y^2 = (x + 1)^2 (2 - x)$.
   Find
   (a) the area of the loop,
   (b) the volume generated when the loop is rotated about the x-axis.

18.6 Calculate the finite area enclosed by the coordinate axis, lines $x = 4$ and $y = 4$ and the curve $xy = 4$.

18.7 Sketch the graph $y = x + \dfrac{3}{x}$ from $0 < x \leqslant 3$.

   Calculate the area of the region bounded by the curve, the x-axis and the lines $x = 1$ and $x = 3$.
   Find the volume of the solid formed by rotating the region through four right angles about Ox, leaving your answer in terms of π.

18.8 Calculate the area of the finite region enclosed between the curves $y = x^3$ and $x = y^2$.
   Find the volume of the solid of revolution formed when the region is rotated through four right angles about the y-axis.

18.4 Outline Solutions to Exercises

18.1 $y = -x^3 + 5x^2 - 6x$, $y' = -3x^2 + 10x - 6$.
At $(2, 0)$ $y' = 2$ equation of tangent $y = 2x - 4$.
At $(3, 0)$ $y' = -3$; equation of tangent $y = -3x + 9$.
Intersect at A $(2.6, 1.2)$.
Area of triangle $ABC = 0.6$ units.

Area under curve $= \displaystyle\int_2^3 (-x^3 + 5x^2 - 6x)\, \mathrm{d}x = \tfrac{5}{12}$.

Required area 0.183.

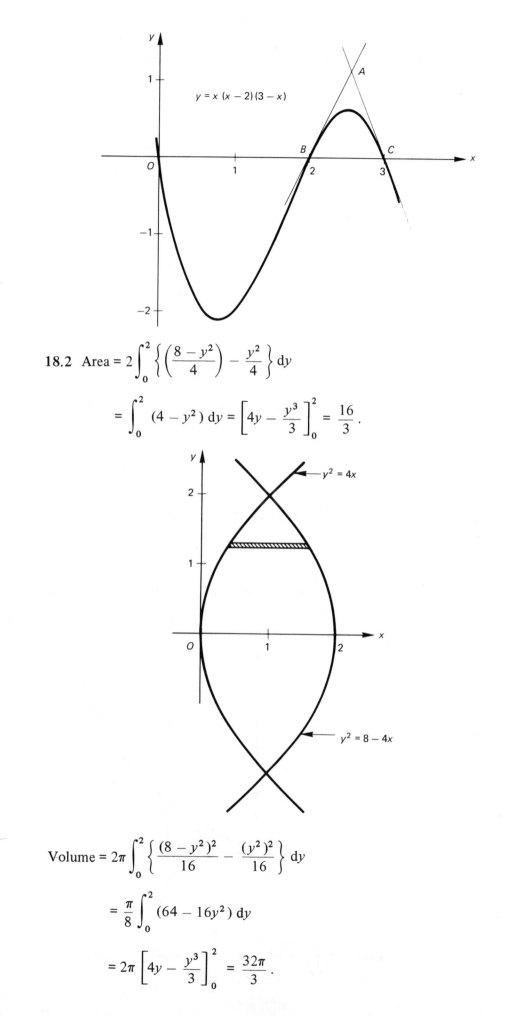

$y = x (x - 2)(3 - x)$

18.2 Area $= 2 \int_0^2 \left\{ \left(\dfrac{8 - y^2}{4} \right) - \dfrac{y^2}{4} \right\} \mathrm{d}y$

$$= \int_0^2 (4 - y^2)\, \mathrm{d}y = \left[4y - \dfrac{y^3}{3} \right]_0^2 = \dfrac{16}{3}\ .$$

$y^2 = 4x$

$y^2 = 8 - 4x$

Volume $= 2\pi \int_0^2 \left\{ \dfrac{(8 - y^2)^2}{16} - \dfrac{(y^2)^2}{16} \right\} \mathrm{d}y$

$$= \dfrac{\pi}{8} \int_0^2 (64 - 16y^2)\, \mathrm{d}y$$

$$= 2\pi \left[4y - \dfrac{y^3}{3} \right]_0^2 = \dfrac{32\pi}{3}\ .$$

217

18.3 Equation of the circle is $x^2 + (y + r)^2 = r^2 \implies x^2 = -y^2 - 2yr$.

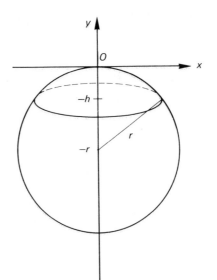

$$\text{Volume of cap} = \pi \int_{-h}^{0} x^2 \, dy = \pi \int_{-h}^{0} (-y^2 - 2yr) \, dy$$

$$= -\pi \left[\frac{y^3}{3} + y^2 r \right]_{-h}^{0} = \pi \left(\frac{-h^3}{3} + h^2 r \right).$$

$$\text{Volume of cap} = \frac{\pi h^2}{3} (3r - h).$$

18.4 (a) $A = \displaystyle\int_{0}^{\pi/3} \tan x \, dx = \left[-\ln |\cos x| \right]_{0}^{\pi/3} = \ln 2 = 0.693$.

$y = \tan x$

$$\text{(b) } V = \pi \int_{0}^{\pi/3} \tan^2 x \, dx = \pi \int_{0}^{\pi/3} (\sec^2 x - 1) \, dx$$

$$= \pi \left[\tan x - x \right]_0^{\pi/3} = \pi \left(\sqrt{3} - \frac{\pi}{3} \right) = 2.15.$$

18.5
$$\text{Area} = 2 \int_{-1}^{2} (x + 1) (2 - x)^{1/2} \, dx.$$

Substitute $u^2 = 2 - x$, $A = -4 \int_{\sqrt{3}}^{0} (3u^2 - u^4) \, du = \dfrac{24\sqrt{3}}{5} = 8.31.$

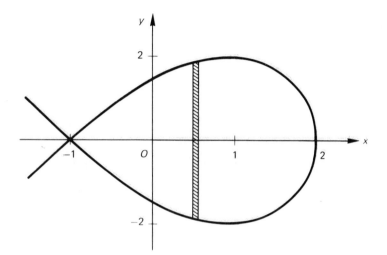

$$\text{Volume} = \pi \int_{-1}^{2} (x + 1)^2 (2 - x) \, dx = \pi \int_{-1}^{2} (2 + 3x - x^3) \, dx$$

$$= \frac{27\pi}{4} = 21.2.$$

18.6

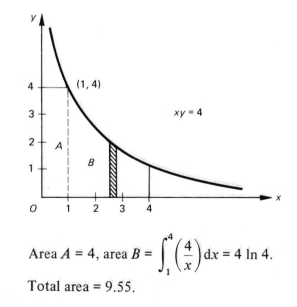

Area $A = 4$, area $B = \displaystyle\int_1^4 \left(\frac{4}{x} \right) dx = 4 \ln 4.$

Total area $= 9.55$.

18.7

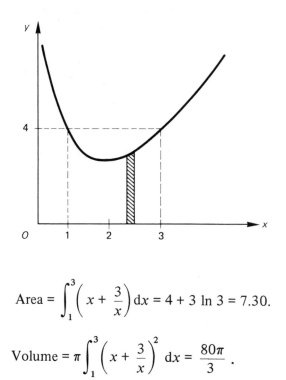

$$\text{Area} = \int_1^3 \left(x + \frac{3}{x} \right) dx = 4 + 3 \ln 3 = 7.30.$$

$$\text{Volume} = \pi \int_1^3 \left(x + \frac{3}{x} \right)^2 dx = \frac{80\pi}{3}.$$

18.8

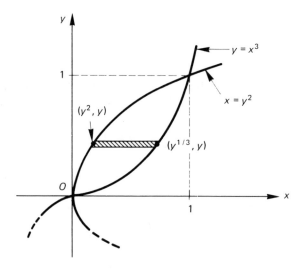

$$\text{Area} = \int_0^1 (y^{1/3} - y^2)\, dy = 5/12.$$

$$\text{Volume about } y\text{-axis} = \pi \int_0^1 (y^{2/3} - y^4)\, dy = \frac{2\pi}{5} = 1.26.$$

Index

*Numbers in italics refer to worked examples

221

222